Jan-Michael Brosi

Slow-Light Photonic Crystal Devices
for High-Speed Optical Signal Processing

AF388435

Karlsruhe Series in Photonics & Communications, Vol. 4
Edited by Prof. J. Leuthold and Prof. W. Freude

Universität Karlsruhe (TH), Institute of High-Frequency and Quantum Electronics (IHQ),
Germany

Slow-Light Photonic Crystal Devices for High-Speed Optical Signal Processing

by
Jan-Michael Brosi

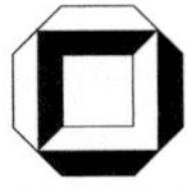

universitätsverlag karlsruhe

Dissertation, Universität Karlsruhe (TH)
Fakultät für Elektrotechnik und Informationstechnik, 2008

Impressum

Universitätsverlag Karlsruhe
c/o Universitätsbibliothek
Straße am Forum 2
D-76131 Karlsruhe
www.uvka.de

Dieses Werk ist unter folgender Creative Commons-Lizenz
lizenziert: http://creativecommons.org/licenses/by-nc-nd/2.0/de/

Universitätsverlag Karlsruhe 2009
Print on Demand

ISSN: 1865-1100
ISBN: 978-3-86644-313-6

Slow-Light Photonic Crystal Devices for High-Speed Optical Signal Processing

Zur Erlangung des akademischen Grades eines

DOKTOR-INGENIEURS

von der Fakultät für
Elektrotechnik und Informationstechnik
der Universität Karlsruhe (TH)

genehmigte

DISSERTATION

von

Jan-Michael Brosi, M. Sc.

geboren in Filderstadt

Tag der mündlichen Prüfung:	18. Juli 2008
Hauptreferent:	Prof. Dr.-Ing. Dr. h. c. Wolfgang Freude
Korreferenten:	Prof. Dr. sc. nat. Jürg Leuthold
	Prof. Dr. rer. nat. Ulrich Lemmer

Contents

Appendix   **119**

Glossary   **145**

Bibliography   **161**

List of Figures

List of Tables

Zusammenfassung (Deutsch)

In dieser Arbeit werden nanophotonische Bauteile auf Basis von photonischen Kristallen mit niedriger Ausbreitungsgeschwindigkeit des Lichtes untersucht. Die Bauteile sollen in der optischen Signalverarbeitung bei hohen Bitraten Anwendung finden. Neuartige Bauteil-Konzepte werden entwickelt, numerische Entwürfe und Simulationen vorgestellt, und Experimente an optischen Bauteilen und an vergrößerten Mikrowellenmodellen ausgeführt.

Durch das rasante Wachstum des Internet-Datenverkehrs und der mobilen Datenkommunikation nehmen die Geschwindigkeits-Anforderungen an Kommunikationsverbindungen ständig zu. Die Kapazität von Kupfer-Leitungen oder Funkverbindungen ist allerdings grundlegend begrenzt, so dass sehr hohe Datenraten nur über kurze Distanzen übertragen werden können. Abhilfe können hier vor allem optische Datenverbindungen schaffen, die um ein Vielfaches größere Kapazitäten zur Verfügung stellen. Tatsächlich ist es zu beobachten dass immer mehr optische Kommunikationstechnik zum Einsatz kommt und dass die optischen Verbindungen immer näher an den Endverbraucher heranrücken. Daneben wird auch schon untersucht, Optik zur Übertragung über kleinste Distanzen, wie z. B. innerhalb von Mikrochips, zu verwenden.

Die Kosten und die Größe von funktionalen optischen Komponenten könnten erheblich reduziert werden, wenn integrierte optische Bauteile eingesetzt werden. Insbesondere die Silizium-Optik ist dabei von großem Interesse, denn die natürlichen Silizium-Rohstoffe sind nahezu unerschöpflich, es können die gleichen Fabrikationsprozesse wie für Mikro-Chips angewendet werden, und es ist prinzipiell möglich sowohl elektronische als auch optische Komponenten auf ein und demselben Chip zu kombinieren. Silizium besitzt zudem einen hohen optischen Brechungsindexkontrast gegnüber Luft, der die benötigten Wellenleiterabmessungen und damit die Chipfläche reduziert. Der hohe Indexkontrast ermöglicht außerdem die Entwicklung von planaren photonischen Kristallen mit einer photonischen Bandlücke.

Ein photonischer Kristall ist eine dielektrische periodische Struktur, die meist künstlich hergestellt wird. In unserem Fall wird eine dünne Silizium-Schicht mit regelmäßig angeordneten zylindrischen Löchern versehen, so dass eine zweidimensionale Periodizität entsteht. Die besondere Eigenschaft eines photonischen Kristalls ist es, dass eine polarisationsabhängige photonische Bandlücke entstehen kann. Das hat zur Folge, dass sich Licht einer bestimmten Polarisation und Frequenz in der Schichtstruktur nicht ausbreiten kann. Wird in einen idealen Kristall ein Defekt-Wellenleiter eingefügt, so kann das Licht darin sehr effektiv geführt werden, und es entstehen zudem Frequenzbereiche mit sehr niedriger Gruppengeschwindigkeit oder sehr hoher chromatischer Dispersion. Diese Eigenschaften wollen wir nutzen, um funktionale Komponenten mit besonderen Eigenschaften für die optische Kommunikationstechnik zu entwerfen.

Zu Beginn der Arbeit zeigen wir die grundlegenden Eigenschaften von photonischen Kristallen auf. Die Wellenausbreitung in photonischen Kristallen wird beschrieben, und dazu werden Banddiagramme eingeführt. Erreicht ein Modus im photonischen Kristall-Wellenleiter den Rand des Banddiagramms, geht die Gruppengeschwindigkeit gegen Null, während die chromatische Dispersion stark anwächst. Dieses Verhalten entsteht durch die Interaktion zweier ursprünglich entgegenlaufender Moden. Wir diskutieren im Weiteren die Anregung von Moden in photonischen Kristallen, und sprechen Verlust-Mechanismen an. Um die Eigenschaften von photonischen Kristallen extern beeinflussen zu können zeigen wir verschiedene Methoden auf.

Anschließend befassen wir uns ausführlich mit dem Entwurf von Wellenleitern in photonischen Kristallen, die eine langsame Lichtausbreitung ermöglichen. Unterschiedliche Wellenleiterstrukturen werden vorgestellt, und Strategien werden entwickelt, um die Gruppengeschwindigkeit oder die chromatische Dispersion durch Veränderung von Strukturparametern einzustellen. Als Beispiele präsentieren wir einen Wellenleiter, der eine kleine Gruppengeschwindigkeit von 4 % der Vakuum-Lichtgeschwindigkeit in einer Bandbreite von 2 THz aufweist, und einen Wellenleiter mit einer hohen negativen chromatischen Dispersion von $-4.5\,\text{ps}/(\text{mm}\,\text{nm})$. Um einen Wellenleiter bei niedriger Gruppengeschwindigkeit effizient mit einem Streifen-Wellenleiter anregen zu können, entwerfen wir einen speziellen Koppel-Taper im photonischen Kristall. Darüber hinaus entwickeln wir eine Methode, um einen Modus mit negativer Gruppengeschwindigkeit anzuregen.

Ungenauigkeiten bei der Herstellung von photonischen Kristallen sind unvermeidlich und führen zu erhöhten Transmissionsverlusten, die die Leistungsfähigkeit maßgeblich beeinträchtigen können. Wir untersuchen durch Simulationen, wie sich die Verluste in verschiedenen Wellenleitern bei langsamem Licht verhalten, und entwickeln Maßnahmen um diese Verluste zu minimieren. Es zeigt sich, dass die geringsten Verluste in einer vertikal symmetrischen Membran-Struktur bei einer maximal möglichen Wellenleiterbreite für einmodiges Verhalten auftreten. Eine vertikale Asymmetrie, bedingt z. B. durch unterschiedliche Materialien im Substrat und in der Deckschicht oder durch Fabrikations-Ungenauigkeiten, führt zu Modenkopplung und damit zu erhöhten Verlusten. Auch ein mehrmodiger Wellenleiter zeigt wesentlich größere Verluste durch Modenkopplung. Bei allen Strukturen wird die Gruppengeschwindigkeit nur unwesentlich durch die Unordnung beeinflusst. Desweiteren untersuchen wir den Zusammenhang zwischen der Größe der Verluste und der Gruppengeschwindigkeit. Im Gegensatz zu früheren Studien können wir aber keinen einfachen Zusammenhang der Verluste mit dem Gruppenbrechungsindex bestätigen.

Die gewonnenen Einsichten und Entwurfsprinzipien werden weiter zur Entwicklung von drei wichtigen Komponenten verwendet: Einem einstellbaren Dispersionskompensator, einer einstellbaren Verzögerungsleitung, und einem schnellen elektro-optischen Modulator.

Wir präsentieren einen Entwurf für einen Dispersionskompensator mit einer regelbaren chromatischen Dispersion zwischen $-19\,\text{ps}/(\text{mm}\,\text{nm})$ und $+7\,\text{ps}/(\text{mm}\,\text{nm})$ in einer 125 GHz großen Bandbreite. Der Kompensator besteht aus zwei hintereinander geschalteten photonische Kristall-Wellenleitern, in denen sich jeweils die Dispersion linear mit der Frequenz ändert. Die Vorzeichen der Steigungen unterscheiden sich dabei, und bei entsprechender Längenanpassung resultiert eine konstante Dispersion. Indem die Charakteristik einer der Sektionen in der Frequenz verschoben wird, z. B. durch Erhitzen, kann die Dispersion eingestellt werden.

Weiterhin entwickeln wir eine optische Verzögerungsleitung, die bei einer Bauteillänge von 1 mm eine variable Verzögerung bis zu 42.2 ps in einer Bandbreite von 125 GHz erreichen kann. Die maximale Verzögerung entspricht dabei der Zeitdauer von 5 Pulsen eines 33 % RZ Daten-Signals bei 40 Gbit/s. Ähnlich wie beim Dispersions-Kompensator werden zwei photonische

Kristall-Wellenleiter hintereinander geschaltet, diesmal aber mit konstanter positiver bzw. konstanter negativer Dispersion. Durch eine Frequenz-Verschiebung einer der beiden Sektionen kann die Verzögerung eingestellt werden.

Der schnelle optische Modulator ist auf Basis eines photonischen Kristalls mit einem schmalen Spalt in der Mitte des Defektwellenleiters aufgebaut. In den Spalt kann ein stark elektrooptisches Material wie z. B. ein Polymer infiltriert werden, in dem eine Phasenverschiebung stattfinden kann. Wir präsentieren einen Modulator-Entwurf mit einer Modulationsbandbreite von 78 GHz und einer Länge von 80 μm bei einer Ansteuerspannung von nur 1 V. Die Bandbreite erlaubt eine Datenrate von 100 Gbit/s. Solche Werte können durch eine hohe Konzentration von sowohl dem optischen als auch dem elektrischen Feld auf den Spalt und durch eine niedrige Gruppengeschwindigkeit, die die nichtlineare Interaktion weiter verstärkt, erzielt werden.

Die Realisierbarkeit von breitbandigen photonischen Kristallen mit langsamer Lichtausbreitung zeigen wir durch Messergebnisse von Wellenleitern, die mit Deep-UV-Lithographie hergestellt wurden. Eine zusätzliche Glas-Deckschicht sorgt für einen vertikal symmetrischen Aufbau. Es wird ein breitbandiger Wellenleiter mit einer Gruppengeschwindigkeit von 3.7 % der Vakuum-Lichtgeschwindigkeit, ein Wellenleiter mit linear variierender Dispersion und einem Dispersions-Minimum von -5 ps / (mm nm), und ein Schlitz-Wellenleiter bei niedriger Gruppengeschwindigkeit vermessen. Die Gruppenverzögerungscharakteristiken stimmen gut mit den Simulationsergebnissen überein, was die Umsetzbarkeit der Entwürfe beweist. Die minimalen Transmissionsverluste liegen für die Wellenleiter zwischen 10 dB / mm und 14 dB / mm. Diese relativ hohen Werte werden vor allem durch die schrägen Seitenwände und durch Oberflächen-Rauigkeiten hervorgerufen. Eine Weiterentwicklung des Fabrikationsprozesses könnte diese Verluste verringern.

Um photonische Kristalle im Experiment mit wesentlich größerer Genauigkeit herstellen und charakterisieren zu können, entwickeln wir ein vergrößertes Mikrowellen-Modell. Dabei werden bei gleicher Brechzahl im Mikrowellenmodell die Strukturen um einen Faktor 20.000 vergrößert und dafür die Betriebsfrequenz um denselben Faktor auf 10 GHz verkleinert. Wir demonstrieren einen photonischen Kristall-Wellenleiter, der umgerechnet auf den optischen Bereich eine Region mit einer Gruppengeschwindigkeit von 4 % der Vakuum-Lichtgeschwindigkeit und eine Region mit einer chromatischen Dispersion von 4 ps / (mm nm) aufweist, jeweils in einer Bandbreite von 1 THz. Die Experimente bestätigen dabei die Simulationen mit der Finite-integration-technique (FIT) sehr gut. Weiterhin können wir am Mikrowellenmodell zeigen, dass zufällige 5-prozentige Störungen der Löcher-Radien die Gruppengeschwindigkeit nicht maßgeblich beeinträchtigen.

Auf Basis der Mikrowellen-Modellexperimente ist es außerdem möglich, verschiedene numerische Werkzeuge zum Entwurf von photonischen Kristall-Wellenleitern zu vergleichen. Wir ermitteln dabei ein akkurates und schnelles Werkzeug für den Entwurf, nämlich die Guided-mode-expansion-Methode (GME). Schließlich wird die Simulation für die Bestimmung der Verluste durch Fabrikations-Ungenauigkeiten überprüft. Dabei stellt sich die FIT Methode als besonders zuverlässig heraus.

Achievements and Limitations

In this work, we study slow-light photonic crystal (PC) waveguides (WGs) in silicon and investigate the potential to realize powerful, cheap and small components for high-speed optical signal processing. Novel device concepts are developed, extensive numerical designs and simulations are performed, and optical experiments as well as scaled microwave experiments are conducted. In this chapter, we summarize our achievements and comment on the limitations of photonic crystal devices.

Group Velocity and Dispersion Engineering: We develop detailed design procedures to engineer the group velocity and chromatic dispersion characteristics of PC-WGs by appropriately changing the structure parameters. Broadband slow light WGs with a group velocity below 4 % of the vacuum speed of light and low chromatic dispersion in a bandwidth of more than 1 THz are designed. PC-WGs with high negative chromatic dispersion well below $-5\,\text{ps}/(\text{mm}\,\text{nm})$ and with spectral ranges of linearly increasing and decreasing dispersion characteristics are developed.

Improved Mode Excitation: We advance a PC coupling taper to significantly improve the excitation of the slow-light PC mode with a strip-WG. Along the taper, selected structure parameters are continuously changed to gradually slow down the PC. The losses per interface can be decreases below $-0.5\,\text{dB}$, and the reflection can be reduced to a value smaller than $-10\,\text{dB}$. Furthermore, a coupling concept is presented that allows exciting a mode with high negative dispersion by injecting power inside a mini-stop band.

Understanding and Minimizing Disorder-Induced Losses: We numerically determine disorder-induced losses of various broadband slow-light PC-WGs. The numerical tools are carefully evaluated, and a loss model is introduced that is needed to analyze the results. We find procedures to minimize losses, and the lowest losses can be found in an air membrane structure with a WG core width as large as possible to still be single-moded. Multi-moded or vertically asymmetric PC-WGs exhibit strong mode-coupling effects that considerably increase losses. Furthermore, we study the dependance of losses on the group velocity characteristics. In contrast to previous findings we cannot verify a simple dependance of the losses on the group velocity.

Tunable Dispersion Compensator: We show a concept for a tunable optical dispersion compensator, and present a design with a dispersion tuning range from $-19\,\text{ps}/(\text{mm}\,\text{nm})$ to $7\,\text{ps}/(\text{mm}\,\text{nm})$ in a 125 GHz bandwidth. The compensator uses two concatenated dispersion-engineered slow-light PC-WGs, where the positive dispersion slope of the first section compensates the negative dispersion slope of the second section.

Tunable Delay Line: We present a tunable delay line with a group delay tuning range of 42.2 ps in a 125 GHz optical bandwidth, while the device length is only 1 mm. The delay range corresponds to 5.1 pulsewidths of a 40 Gbit/s 33 % RZ data signal. The delay line is realized by two concatenated PC-WGs having constant dispersion values of opposite sign.

Electro-Optic Modulator: We propose a high-speed modulator with low drive voltage based on a slow-light PC-WG. A highly nonlinear organic material is infiltrated into a narrow slot in the WG center, where both the optical and the electric fields are strongly confined. For a design with negligible first-order chromatic dispersion in an optical bandwidth of 1 THz, we predict a modulation bandwidth of 78 GHz and a length of about 80 μm at a drive voltage amplitude of 1 V. This allows transmission at 100 Gbit/s. We estimate that the modulation bandwidth is limited by the walk-off between the optical and electrical signals, while $RC-$effects do not play a role. This work has been published in a journal article, [1].

Optical Experiments: We design and characterize slow-light PC-WGs fabricated with deep-UV lithography. A broadband slow-light WG with a group velocity of 3.7 % of the vacuum speed of light, a WG with linearly varying chromatic dispersion and a dispersion minimum of -5 ps $/$(mm nm) and a slow-light slot-WG are realized. The measured group delays for all three PC-WGs agree with the simulated characteristics thus verifying the designs. The minimum transmission losses are between 10 dB $/$ mm and 14 dB $/$ mm.

Microwave Model Experiments: We develop an enlarged microwave model for the experimental proof-of-concept for novel photonic devices. In comparison to optical experiments, the microwave structures can be fabricated to a much higher accuracy and measurement equipment is much more precise at frequencies near 10 GHz, which also allows verifying numerical tools. We demonstrate a broadband slow-light PC-WG that exhibits a region with a group velocity of 3.4 % of the vacuum speed of light in a 1.2 THz optical bandwidth and a region with an optical chromatic dispersion of 3.9 ps $/$ (mm nm) in a 1.1 THz bandwidth. Simulations with the finite integration technique (FIT) predict the device behavior correctly. We show by experiments that a 5 % disorder in the hole radii does not significantly impair the group velocity characteristics thus proving the feasibility of a realistic device. Parts of this work have been published in a journal article, [2].

Numerics Verification: Various numerical tools for PC-WG band calculations and for complete device simulations are verified and their suitability is evaluated. For the design of slow-light PC-WGs, we prove that the guided-mode expansion (GME) method is an accurate and very fast tool. However, for disorder simulations, the results from GME are distorted due to the approximations used in the method, while the much slower FIT models the influence of disorder correctly. Parts of this work have been published in a journal article, [2].

While silicon slow-light PC-WGs have unique properties and are very promising, we also need to discuss the existing limitations.

First, the disorder-induced losses of PCs that are caused by fabrication imperfections can degrade the device performance, especially at low group velocity. For short devices with PC-WG lengths well below 1 mm, like with the modulator, the losses can be tolerated. But for devices with lengths in the order of 1 mm or larger, like with the dispersion compensator or the delay line, losses might become considerable. In addition, vertical asymmetries, either intended or caused by fabrication imperfections, can lead to mode coupling effects that induce unwanted transmission dips. On the other hand, fabrication processes are continually improved, which decreases disorder-induced losses.

Second, slow-light PC-WGs, like all other slow-light media, exhibit fundamental bandwidth limitations, and the lower the group velocity or the higher the chromatic dispersion is, the smaller is the achievable bandwidth. For a broadband slow-light PC-WG with a group velocity of 4% of the vacuum speed of light, the spectral range of low chromatic dispersion cannod exceed much more than 2 THz or 16 nm near the telecommunication wavelength of $1.55\,\mu$m. Furthermore, the spectral characteristics of PC-WGs are not periodic. Thus, in communication systems with wavelength division multiplexing, a slow-light PC-WG device can only process one or few wavelength channels, whereas many channels would require several PC devices.

Third, the coupling from a fiber to the small silicon PC device is not trivial and leads to losses. Furthermore, only the transverse-electric (TE) mode in the PC should be excited, as the PC is highly polarization dependent. The best reported coupling efficiencies from fibers to silicon WGs are around 70% per facet, and efficient polarization controllers or splitters exist. However, it would be desirable to find simpler and more efficient methods.

Conclusion: Photonic crystal waveguides offer great opportunities in specific areas, especially for applications where tailoring the dispersion characteristics is of interest.

Summary

With the fast evolution of wired and mobile communications, regional as well as world-wide data traffic is vastly increasing. The growth of new high-speed applications like video streaming and video on demand and the continuous increase of information services and business activities in the Internet have increased the need for high-speed data links. While the bandwidth of wireless and copper-based data channels is limited, optical links offer the required high bandwidth, and the trend is towards the deployment of optics up to the end-user or even inside homes and company buildings. In optical data networks of different size and complexity, various optical components are needed, like e. g. lasers, modulators, dispersion compensators, delay lines, receivers, routers or regenerators; all of these components can be categorized as devices for optical signal processing. In recent years, integrated optical components have been extensively investigated, as they have the potential to considerably reduce the size and cost of functional optical devices. The materials for these devices can be glasses, polymers or semiconductors. Among these materials, Silicon-on-Insulator (SOI) is of particular interest, as the natural material resources are almost unlimited, optical SOI devices can be fabricated with the same mature processes that are used for microchips and are suitable for cheap mass-production, and electrical and optical devices can even be combined on the same SOI chip. The index contrast of silicon towards the cladding materials air or glass is very high, which confines the optical fields very strongly to the waveguides (WGs) and reduces the required WG size. Furthermore, the high index contrast of SOI allows to build photonic crystals (PC), which are artificial periodic structures with unique optical properties.

A silicon photonic crystal can be designed to have a complete photonic bandgap (PBG) for one polarization, which means that the propagation of light is inhibited in a certain frequency range. If a line defect WG is formed in such a PC, the light in this PC-WG can be very effectively guided. Furthermore, the PC-WG modes exhibit regions of very low group velocity and very high chromatic dispersion, and the PC structure offers many degrees of freedom to engineer the dispersive properties. It is e.g. possible to achieve an optical group velocity of only 4% of the vacuum speed of light, or a chromatic dispersion that is larger by a factor of a million in comparison to the dispersion of a single mode fiber (SMF). At the same time, bandwidths in the order of 1 THz or higher can be obtained, which are required for transmitting high-speed optical data signals. The very special group velocity and dispersion characteristics of PC-WGs allow developing a new class of compact functional integrated devices for high-speed optical signal processing.

In this work, we study the exceptional properties of PC-WG and develop design procedures for PC devices with low group velocity or high dispersion, while also taking the limitations by imperfect fabrication processes into account. By using the gained basic understanding, we develop concrete designs for a variable optical delay line, for a residual dispersion compensator, and for a fast and compact electro-optic modulator. In addition to the numerical designs and device simulations, we perform optical experiments and microwave model experiments, which proof the validity of the numerics and demonstrate the feasibility of the proposed devices.

In Chapter 1 of this thesis, we review the fundamental properties of PCs. The wave propagation in strip-WG and two-dimensional (2D) PCs is discussed, and band diagrams are introduced that characterize the PC properties. Bulk PCs without defect reveal a complete PBG for transverse electric (TE) fields, while inside a defect-WG guided modes appear, which have a group velocity that approaches zero and a chromatic dispersion that grows to infinity as the band-edge is reached. The special modal behavior in PCs results from coupling mechanisms between counter-propagating waves, and the emergence of structure asymmetries can lead to additional coupling phenomena. The excitation of the slow-light PC modes is addressed, and loss mechanisms are discussed. To obtain functional components, the structure behavior needs to be externally controlled, and we present different methods to tune PCs.

In Chapter 2, we concentrate on slow-light PC-WGs. The first part of the chapter is devoted to the WG design, Section 2.1. We discuss three different WG geometries and their characteristics, and we work out strategies to optimize the optical behavior of PC-WG and to engineer their group velocity and chromatic dispersion by changing the structure parameters. In particular, we present a broadband slow-light PC-WG having a group velocity of 4% of the vacuum speed of light in a 2 THz bandwidth, and a PC-WG with linearly varying chromatic dispersion and a minimum dispersion value of $-4.5\,\mathrm{ps}/(\mathrm{mm\,nm})$. For the excitation of slow-light PC-modes with a conventional strip-WG, we develop an efficient method that employs a carefully designed PC-taper and leads to significantly increased coupling. In addition, we present a new method that allows exciting a counter-propagating mode. In the second part of the chapter, Section 2.2, fabrication-induced losses are treated, which represent a major limitation of the slow-light PC-WGs. We determine the disorder-induced losses of different broadband slow-light PC-WGs with numerical simulations, and develop procedures to minimize losses. We find that a vertically symmetric structure like an air-membrane structure is best suited for low-loss propagation, as a vertical asymmetry leads to TE-TM coupling. Increasing the WG core to a value as large as possible decreases losses, however a multi-moded behavior needs to be avoided. A lower cladding index and a larger slab thickness can increase the operation range below the light line. The group velocity characteristics of the slow-light WGs are only slightly influenced by the disorder. We also relate the losses to the group velocity, and in contrast to previous findings we cannot verify a linear or quadratic dependence of the losses on the group index.

In Chapter 3, we apply the gained knowledge about slow-light PC-WGs to develop three novel optical devices: A variable chromatic dispersion compensator, a variable optical delay line, and a fast electro-optic modulator. An exact chromatic dispersion compensation for optical links is important to reduce bit errors especially at high bit-rates, and affordable and small components are needed for the variable compensation of residual dispersion, e. g. after a fixed length of dispersion compensating fiber. In Section 3.1 we present the design of a tunable dispersion compensator with a dispersion tuning range from $-19\,\mathrm{ps}/(\mathrm{mm\,nm})$ to $+7\,\mathrm{ps}/(\mathrm{mm\,nm})$ in a broad 125 GHz bandwidth. The compensator is based on two concatenated slow-light PC-WG sections, where the first section has a linearly increasing dispersion with frequency, while the second section shows a linearly decreasing dispersion. If the dispersion slopes of both sections have opposite sign, the resulting dispersion is constant in the operating bandwidth, and the dispersion value can be changed by shifting the characteristics of one section in frequency. In

Section 3.2, we suggest a compact PC-device that can realize a tunable optical delay, which is applicable in next-generation optical networks e.g. for data synchronization, processing or storage, and can also be used to realize true time-delays e. g. in RADAR applications. We present a design with a large tuning range of $\Delta t_g = 42\,\text{ps}$ in an optical bandwidth of 125 GHz at a device length of only 1 mm. In this device, a 40 Gbit/s optical 33 % RZ signal can be shifted in time between 0 and 5 pulse widths. The delay line is, similarly to the tunable dispersion compensator, based on two concatenated PC-WG sections, where now the first section has a constant positive dispersion, whereas the second section exhibits a constant negative dispersion, and the lengths of both sections are chosen such that the total dispersion vanishes. If again one of the sections is tuned to shift its characteristics in frequency, the delay can be adjusted. Finally, in Section 3.3 we present an electro-optic silicon-based modulator with a low drive voltage, which has the potential to significantly reduce costs and increase the performance of fast communication systems. The modulator has a modulation bandwidth as high as 78 GHz at a low drive voltage amplitude of 1 V and a length of only $80\,\mu\text{m}$. Such a modulator allows 100 Gbit/s transmission and can be achieved by infiltrating an organic material with high electro-optic coefficient in a narrow slot in the center of a slow-light PC-WG. Both the optical and the electrical fields are strongly confined to the slot, and the low optical group velocity increases the nonlinear interaction.

Chapter 4 is devoted to prove the feasibility of the presented slow-light structures with optical and microwave experiments, and to verify the numerical modeling tools. In Section 4.1, we show optical measurement results of silicon structures fabricated with deep-UV-lithography. To obtain a vertically symmetric slab structure, a glass cladding is deposited on top. For a basic strip-WG, we determine the transmission losses to be around 1 dB / mm. The fabricated and measured slow-light PC-WGs are a broadband slow-light WG with a group velocity of 3.7 % of the vacuum speed of light, a PC-WG with a linearly varying chromatic dispersion and a dispersion minimum of $-5\,\text{ps} / (\text{mm\,nm})$, and a slow-light slot PC-WG with a slot width of 180 nm. For all three PC-WGs, the measured group delay characteristics agree with the simulated group delay, which indicate that the group velocity and dispersion actually behave as designed. The minimum transmission losses are 14 dB / mm for the broadband slow-light WG, 11.2 dB / mm for the WG with linearly varying dispersion, and 10 dB / mm for the slot WG. The losses are mainly caused by non-vertical sidewalls and surface roughness, and can be decreased by further optimizing the fabrication process. To be able to fabricate and characterize the properties of slow-light PC-WGs with much higher precision, we have developed an upscaled experimental method in the microwave regime, which is presented in Section 4.2. The device dimensions are enlarged by a factor of 20,000, while the frequency is reduced by the same factor. Microstrip-fed slot antennas feed the microwave strip WGs, and continuous wave measurements, broadband pulse transmission experiments and field distribution measurements are performed. We demonstrate a broadband slow-light WG that exhibits a region with a low group velocity of 4 % of the vacuum speed of light and a region with a high chromatic dispersion of 4 ps / (mm nm), both in a 1 THz bandwidth. The experiments verify the predictive power of device simulations with the finite integration technique (FIT). In addition, we quantify by experiments that a random disorder of the photonic crystal's hole radii by 5 %, which can be caused by fabrication imperfections, does not degrade the group velocity behavior significantly. Based on the microwave experiments, several numerical band calculation methods to design PC-WGs are benchmarked, see Section 4.3. We find that the guided mode expansion (GME) method is a fast and accurate semi-analytical design tool for PC-WGs, and that also the FIT technique is well suited to model PC devices, however with longer simulation times compared to GME. To also gain trust in the numerical loss predictions for PC-WGs with disorder, we compare results from different methods, and find out that the FIT method is also well suited for disorder simulations.

In the Appendix, mathematical derivations are presented in more detail, and measurement and simulation techniques are explained. Appendix A.1 states the used mathematical transformations and the signal representation. Appendix A.2 gives a short overview over the different numerical methods used in this work. In Appendix A.3, the orthogonality relation for PC modes and the connection between the group velocity and the mode fields are derived. The Through-Reflect-Line (TRL) calibration method used for the microwave experiments and for the disorder simulations is reviewed and adapted in Appendix A.4. The coupling of counter-propagating modes that leads to gaps in the band diagram is treated in Appendix A.5, and the loss model based on coupled power equations as used for the disorder simulations is developed and solved in Appendix A.6. The derivations needed for estimating the modulator field interaction factor and the bandwidth limitations are presented in Appendix A.7. Details about the microwave measurement techniques used for the scaled microwave experiments are given in Appendix A.8.

Chapter 1

Fundamentals

The simplest form of a photonic crystal is a one-dimensional thin film stack, where two different materials with a thickness of a quarter wavelength are alternatively arranged, which leads to frequencies that are transmitted while others are reflected. While the concept of a thin-film stack has already existed for more than a century, it was only much later that the idea to form a periodic lattice in two or three dimensions has appeared in order to create artificial materials that exhibit frequency ranges in which no wave can propagate. The first theoretical papers on two-dimensional (2D) or three-dimensional (3D) PCs were published in 1987 by Yablonovitch [3] and John [4], followed by experimental evidence in the microwave regime by Yablonovitch [5], [6] a few years later. In the following time, a lot of interest arose in the topic leading to intense theoretical and experimental research activities. The attraction of PCs relies on two basic properties: First, a complete band gap can be formed that inhibits wave propagation, while the introduction of a defect can lead to very effective light confinement or guiding. Second, PCs can exhibit unique refractive and dispersive characteristics which can be used for a variety of applications. In this chapter, the fundamental properties of PCs needed for this work are reviewed.

This chapter is structured as follows: In Section 1.1, the wave propagation in dielectric media and in translationally invariant waveguides is discussed. In Section 1.2, the properties of 2D PCs are explained, and the focus is especially on PC-WGs in a slab of high-index material. Section 1.3 is dedicated to group velocity and chromatic dispersion, and to the special characteristics that can be found in PCs. In Section 1.4, the phenomena of so-called mode-gaps and mini-stop bands are explained, and Section 1.5 addresses the issue of mode excitation in PC-WGs. An introduction into loss mechanisms in PCs is given in Section 1.6, and finally, three possible tuning mechanisms to dynamically change the material properties of silicon-based PCs are presented in Section 1.7.

1.1   Wave Propagation in Dielectric Media

1.1.1   Maxwell's Equations and Scaling Laws

The propagation of waves in a source-free, nonmagnetic dielectric linear isotropic medium is governed by *Maxwell's equations* in the following form [7], [8]:

$$\nabla \times \mathbf{E} = -\mu_0 \frac{\partial \mathbf{H}}{\partial t}, \tag{1.1}$$

$$\nabla \times \mathbf{H} = \varepsilon_0 \varepsilon_r \frac{\partial \mathbf{E}}{\partial t}, \tag{1.2}$$

$$\nabla \cdot \mathbf{D} = 0, \tag{1.3}$$

$$\nabla \cdot \mathbf{B} = 0. \tag{1.4}$$

The electric displacement $\mathbf{D}$ is related to the electric field $\mathbf{E}$ by

$$\mathbf{D} = \varepsilon_0 \varepsilon_r \mathbf{E}, \tag{1.5}$$

where $\varepsilon_0 = 8.85419 \times 10^{-12}\,\mathrm{As}/(\mathrm{Vm})$ is the permittivity of vacuum and ε_r the relative permittivity. In general, the relative permittivity ε_r is a tensor, but in this work we mainly consider a scalar permittivity. The magnetic flux density $\mathbf{B}$ of a nonmagnetic material is related to the magnetic field $\mathbf{H}$ by

$$\mathbf{B} = \mu_0 \mathbf{H}, \tag{1.6}$$

with the permeability of vacuum $\mu_0 = 1.25664 \times 10^{-6}\,\mathrm{Vs}/(\mathrm{Am})$. In general, the vector fields $\mathbf{D}$, $\mathbf{E}$, $\mathbf{B}$ and $\mathbf{H}$ and the relative permittivity ε_r are functions of time t and space coordinate $\mathbf{r} = (x, y, z)$. For harmonic solutions with a sinusoidal time dependance, the fields can be written in complex form as:

$$\mathbf{E}(t, \mathbf{r}) = \mathbf{E}(\mathbf{r})\,\mathrm{e}^{\mathrm{j}\omega t}, \tag{1.7}$$

$$\mathbf{H}(t, \mathbf{r}) = \mathbf{H}(\mathbf{r})\,\mathrm{e}^{\mathrm{j}\omega t}. \tag{1.8}$$

The physical fields can be obtained from the complex fields by taking the real part. The angular frequency ω and the frequency f are related by $\omega = 2\pi f$. If the time-dependance Eqs. (1.7), (1.8) is substituted in the first two lines of Maxwell's equations Eqs. (1.1), (1.2), a modified form is obtained:

$$\nabla \times \mathbf{E}(\mathbf{r}) = -\mathrm{j}\omega\mu_0 \mathbf{H}(\mathbf{r}), \tag{1.9}$$

$$\nabla \times \mathbf{H}(\mathbf{r}) = \mathrm{j}\omega\varepsilon_0\varepsilon_r \mathbf{E}(\mathbf{r}). \tag{1.10}$$

For a dielectric material, the material properties can also be characterized by the refractive index $n = \sqrt{\varepsilon_r}$ instead of the relative permittivity ε_r. If we divide Eq. (1.10) by ε_r, apply the operator $\nabla\times$ from the left hand side, and use Eq. (1.9) to eliminate $\mathbf{E}$, an eigenvalue equation for the magnetic fields can be found:

$$\left(\nabla \times \left(\frac{1}{\varepsilon_r(\mathbf{r})}\nabla \times\right)\right)\mathbf{H}(\mathbf{r}) = \left(\frac{\omega}{c}\right)^2 \mathbf{H}(\mathbf{r}). \tag{1.11}$$

The vacuum speed of light is $c = 1/\sqrt{\varepsilon_0 \mu_0}$. If the magnetic fields are determined from Eq. (1.11), the electric fields can be provided by using Eq. (1.10).

A general feature of wave propagation in dielectric media is that there is no fundamental length scale, and also no fundamental value of the dielectric constant. This is expressed in the *scaling laws of Maxwell's equations* [9]. If the dielectric structure is enlarged or diminished by a scaling factor s to $\varepsilon_r'(\mathbf{r}) = \varepsilon_r(\mathbf{r}/s)$ using $\mathbf{r}' = s\mathbf{r}$ and $\nabla' = \nabla/s$, the eigenvalue equation (1.11) can be rewritten as:

$$\left(\nabla' \times \left(\frac{1}{\varepsilon_r'(\mathbf{r}')} \nabla' \times \right) \right) \mathbf{H}(\mathbf{r}'/s) = \left(\frac{\omega}{sc} \right)^2 \mathbf{H}(\mathbf{r}'/s) . \tag{1.12}$$

This is just Eq. (1.11) again, but with the field solution $\mathbf{H}'(\mathbf{r}') = \mathbf{H}(\mathbf{r}'/s)$ and the frequency $\omega' = \omega/s$. If the length scale is changed by a factor s, the field solution and its frequency are scaled by this same factor. The solution at one length scale gives the solution at all other length scales. Similarly, if the dielectric permittivity is scaled in the whole domain by a factor s^2 to $\varepsilon_r'(\mathbf{r}) = \varepsilon_r(\mathbf{r})/s^2$, Eq. (1.11) is rewritten as:

$$\left(\nabla \times \left(\frac{1}{\varepsilon_r'(\mathbf{r})} \nabla \times \right) \right) \mathbf{H}(\mathbf{r}) = \left(\frac{s\omega}{c} \right)^2 \mathbf{H}(\mathbf{r}) . \tag{1.13}$$

The field solution remains unchanged, but the frequency is scaled to $\omega' = s\omega$.

1.1.2 Modes of Translational Invariant Waveguides

A conventional optical waveguide is a dielectric structure that is translational invariant in one direction and can guide the light along this direction. A mode of a WG is a field distribution that does not change its shape while it propagates. The structure of a dielectric strip WG is shown in Fig. 1.1. The refractive index n_{core} of the WG core needs to be larger than the refractive indexes of the substrate n_{sub} and of the cover material n_{cover} in order to obtain a guided mode in the core. The direction of the WG is along the z-axis. A mode at frequency ω with a harmonic time dependance $e^{j\omega t}$ as in Eqs. (1.7) and (1.8) has to satisfy the Maxwell's equations (1.9), (1.10) and can be written in the form:

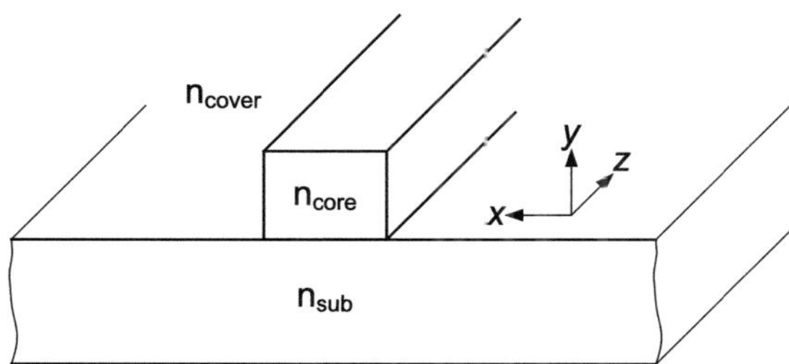

Fig. 1.1. Schematic of a strip waveguide. The refractive indices of the waveguide core, the substrate and the cover are denoted as n_{core}, n_{sub}, and n_{cover}, respectively.

$$\mathbf{E}(x,y,z) = \hat{\mathbf{E}}(x,y)\,\mathrm{e}^{-\mathrm{j}\beta z}, \tag{1.14}$$

$$\mathbf{H}(x,y,z) = \hat{\mathbf{H}}(x,y)\,\mathrm{e}^{-\mathrm{j}\beta z}. \tag{1.15}$$

The propagation constant β and the transverse field profile $\hat{\mathbf{E}}, \hat{\mathbf{H}}$ are functions of the frequency ω. The dependance of the propagation constant with frequency $\beta(\omega)$ is referred to as the dispersion relation of the WG mode. For a positive propagation constant, the phase fronts propagates in positive z-direction, for a negative propagation constant in negative z-direction. In many cases, the effective refractive index n_{eff} is displayed as a function of frequency instead of the propagation constant β, and the relation between the two quantities is:

$$\beta = n_{\mathrm{eff}}\,\frac{\omega}{c}. \tag{1.16}$$

Modes with a dominant field component E_x are called (quasi)-transverse electric (TE) modes, whereas modes with a dominant electric field component E_y are called (quasi)-transverse magnetic (TM) modes. A mode is guided if the effective refractive index exceeds the substrate index, and for a lower effective index the mode becomes lossy and leaks into the substrate.

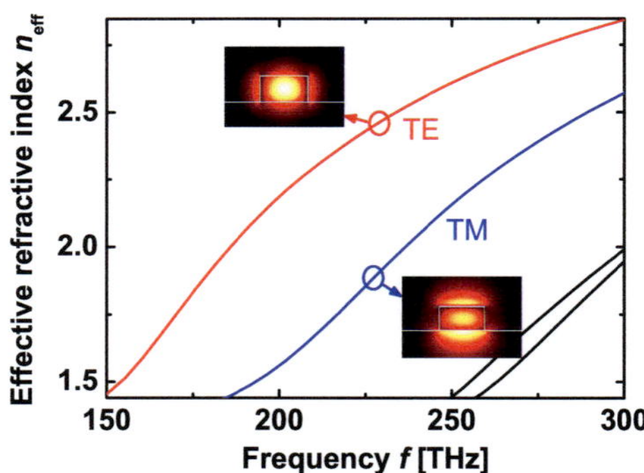

Fig. 1.2. Dispersion relation of a SOI waveguide. The WG is 220 nm in height and 400 nm in width. The distributions of the dominant electric field components for the fundamental TE and TM modes are displayed as insets. Higher order modes only appear for frequencies above 250 THz.

In Fig. 1.2, the dispersion relation of a strip WG in the SOI material system is shown. The refractive indices of Silicon and glass at optical frequencies near 193 THz are $n_{\mathrm{core}} = 3.48$, $n_{\mathrm{sub}} = 1.44$, respectively, and the cladding material is air with $n_{\mathrm{cladding}} = 1$. The strip is chosen to have a thickness of 220 nm and a width of 400 nm. At an operating frequency near 193 THz, the WG guides the fundamental TE and TM modes, but the TM mode is near cutoff and the fields extend largely into the cladding. The typical field distributions of the TE and TM modes are displayed as insets, where the dominant electric field component is plotted (E_x for the TE-mode and E_y for the TM-mode). Higher-order modes only appear for frequencies above 250 THz.

For a fixed frequency, the field distribution inside the WG can be represented as a superposition of guided and radiative modes, [10]. For many purposes, it is sufficient to only consider the guided modes in the expansion. The *orthogonality relation* for two guided modes $\mathbf{E}_{p,q}$, $\mathbf{H}_{p,q}$ subscripted with p and q, respectively, reads [10]:

$$\frac{1}{4}\iint \left(\hat{\mathbf{E}}_q \times \hat{\mathbf{H}}_p^* - \hat{\mathbf{H}}_q \times \hat{\mathbf{E}}_p^*\right)\, \mathbf{e}_z dA = \delta_{p,q} P_p\,. \tag{1.17}$$

The surface integration is performed over a cross-section perpendicular to the propagation direction, $\delta_{p,q}$ is the Kronecker-Delta, and P_p is the cross-section power of mode p. If $p = q$ is set in Eq. 1.17, the *power of mode p* is determined:

$$P_p = \frac{1}{2}\iint \Re\left(\hat{\mathbf{E}}_p \times \hat{\mathbf{H}}_p^*\right)\cdot \mathbf{e}_z dA\,. \tag{1.18}$$

1.2   2D Photonic Crystals

A photonic crystal is in general a periodic dielectric structure in one, two or three dimensions [9]. The focus in this work is especially on two-dimensional PCs formed by introducing a hexagonal lattice of cylindrical holes with low index in a slab of a high-index material [11], which can exhibit a complete bandgap for TE polarization and can be realized in SOI with standard fabrication processes. By omitting a row of holes in a perfect lattice, a line defect WG [12], [13] is formed in the PC that can guide the light efficiently. The structure of a PC and a PC-WG are schematically shown in Fig. 1.3.

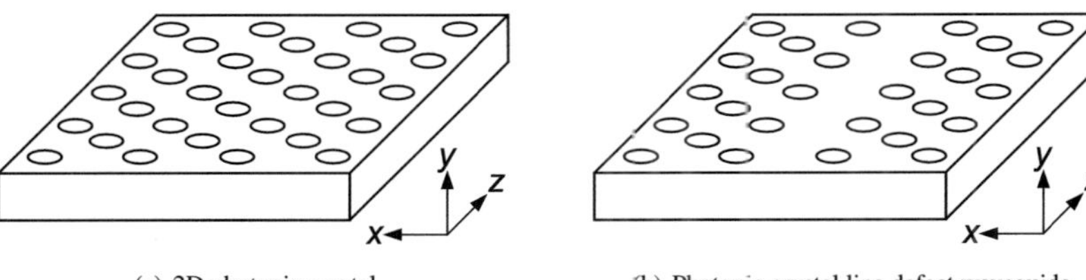

(a) 2D photonic crystal.　　　　　　　　　　　(b) Photonic crystal line defect waveguide.

Fig. 1.3. Schematic of a photonic crystal (PC) and a PC line defect waveguide. (a) The PC consists of a hexagonal lattice of cylindrical air holes in a slab of high-index material. (b) The PC-WG is formed by omitting a row of holes in a perfect PC lattice.

1.2.1   Bulk Photonic Crystal

In this work, we refer to an ideal PC without any defect as bulk photonic crystal, Fig. 1.3(a). The structure of a PC does not have a continuous translational symmetry like a strip-WG has in z-direction, Fig. 1.1, but a discrete translational symmetry in the xz-plane. The PC lattice can be represented using two primitive lattice vectors $\mathbf{a}_1$, $\mathbf{a}_2$ of length a, where a is the lattice constant of the PC, Fig. 1.4(a). A possible set of primitive lattice vectors is $\mathbf{a}_1 = a\,\mathbf{e}_z$ and $\mathbf{a}_2 = -\frac{1}{2}\sqrt{3}\,a\,\mathbf{e}_x + \frac{1}{2}a\,\mathbf{e}_z$. The position $\mathbf{R}$ of each hole center in the lattice can be represented by a linear combination of the primitive lattice vectors with integer coefficients l_1 and l_2, $\mathbf{R} = l_1\,\mathbf{a}_1 + l_2\,\mathbf{a}_2$. The symmetry of the dielectric structure can thus be represented as $\varepsilon_r(\mathbf{r}+\mathbf{R}) = \varepsilon_r(\mathbf{r})$. As the dielectric structure is symmetric, also the eigenfunctions must exhibit the same symmetry. The eigenfunctions with two-dimensional translational periodicity can be expanded into plane waves $e^{-j\mathbf{G}\cdot\mathbf{r}}$ having propagation direction $\mathbf{G}$. This expansion corresponds to a two-dimensional Fourier transform of the eigenfunctions. The vectors $\mathbf{G}$ of the plane-wave expansion are called the reciprocal lattice vectors, and they form the reciprocal lattice, Fig. 1.4(b). As in the direct lattice, all vectors in the reciprocal lattice $\mathbf{G}$ are a linear combination of two primitive lattice vectors $\mathbf{b}_1$, $\mathbf{b}_2$ with integer coefficients. A possible set of reciprocal lattice vectors is $\mathbf{b}_1 = -\frac{4\pi}{\sqrt{3}a}\,\mathbf{e}_x$ and $\mathbf{b}_2 = -\frac{2\pi}{\sqrt{3}a}\,\mathbf{e}_x + \frac{2\pi}{a}\,\mathbf{e}_z$. The eigenfunctions of the structure can be classified by a wave vector $\mathbf{k}$. *Bloch's theorem* states that the eigenfunctions (or modes) can be represented in the following form [9]:

$$\mathbf{H}(\mathbf{r}) = e^{-j\mathbf{k}\cdot\mathbf{r}}\,\mathbf{u}(\mathbf{r}) \quad \text{with } \mathbf{u}(\mathbf{r}) = \mathbf{u}(\mathbf{r}+\mathbf{R})\,. \tag{1.19}$$

The mode field $\mathbf{H}$ for a given wave vector $\mathbf{k}$ has a periodic part $\mathbf{u}(\mathbf{r})$ with periodicity in the lattice vectors $\mathbf{R}$, and has a phase factor $e^{-j\mathbf{k}\cdot\mathbf{r}}$. The same law holds for the $\mathbf{E}$-field.

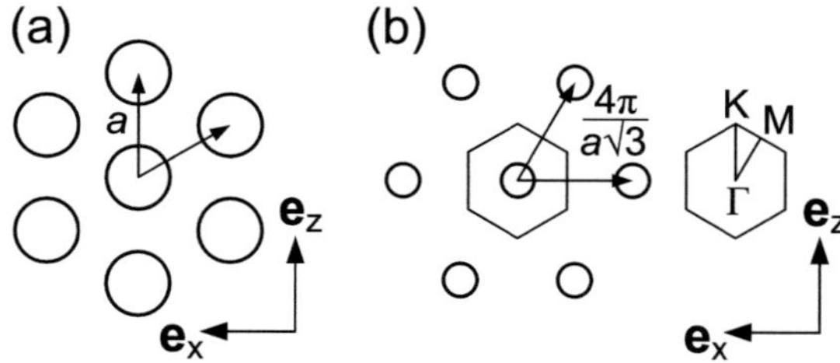

Fig. 1.4. Lattice of a two-dimensional photonic crystal. (a) Direct lattice and (b) reciprocal lattice with first Brillouin zone (hexagon). The lattice constant is a. The characteristic points in the Brillouin zone are denoted as Γ, K and M.

Another important property that can be deduced from Bloch's theorem is that modes with wave vector $\mathbf{k}$ and with wave vector $\mathbf{k}+\mathbf{G}$ are identical if $\mathbf{G}$ is an arbitrary reciprocal lattice vector. This means that we can restrict ourselves to analyze the modes in a certain domain in $\mathbf{k}$-space, as if we pick a mode with $\mathbf{k}$-vector outside of this domain, we can find a $\mathbf{k}$-vector inside the domain which refers to exactly the same mode. The smallest possible domain in $\mathbf{k}$-space that can address all modes is the so-called *first Brillouin zone*, which is the area confined by the hexagon in Fig. 1.4(b). As the structure has also rotational and mirror symmetries, the domain can be further reduced to the area confined by the triangle in the Brillouin zone. It has been found that in order to characterize the dispersive properties of the PC, it is enough to determine the eigenvalues ω of the eigenfunctions $\mathbf{H}$ for the $\mathbf{k}$-vectors that lie on the edges of the triangle with vertices Γ, M and K. This is graphically represented in the *band diagram*. The typical band diagram of a membrane PC for both TE and TM polarization is shown in Fig. 1.5. It has been calculated for a silicon slab surrounded by air. The height of the slab is $h = 0.55\,a$ and the

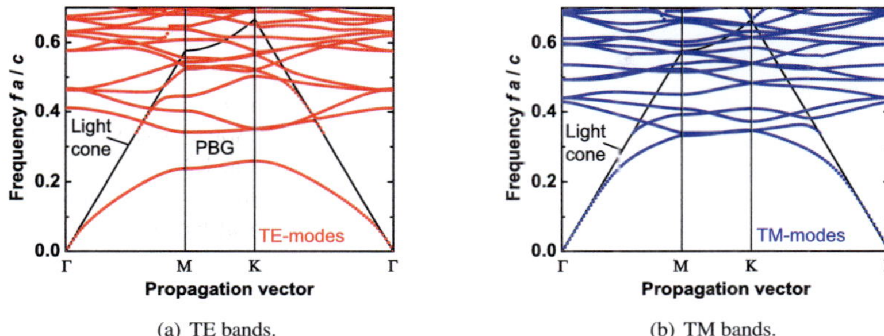

(a) TE bands.   (b) TM bands.

Fig. 1.5. Band diagram of a bulk PC for TE and TM-polarized modes. (a) For the TE-modes, a frequency region without any mode can be observed, this is the photonic bandgap (PBG). (b) For the TM-modes, no photonic bandgap exists. In the region below the light cone, the modes are guided in the slab. The PC lattice constant is a.

radius of the holes is $r = 0.3\,a$, where the lattice constant a is scalable due to the scaling laws, Eq. 1.12. The TE band diagram, Fig. 1.5(a), clearly shows a considerably large frequency range without any modes, this is the photonic bandgap (PBG). For the TM band diagram, Fig. 1.5(b), there is no PBG. The light cone represents the dispersion of the substrate material, which is in the given case air. If a mode lies below the light cone, the mode is confined to the material slab, and for a perfect structure only material losses occur. If a mode lies above the light cone, the mode can leak into the surrounding space, which leads to increased losses.

1.2.2 Photonic Crystal Waveguide

The simplest form of a WG inside a PC can be obtained by removing a row of holes in a bulk PC, Fig. 1.3(b). The structure is not any longer translational symmetric in two dimensions, but only in one dimension along the WG axis, which is the z-axis. The wave vector $\mathbf{k}$ is oriented along z-direction, $\mathbf{k} = k\,\mathbf{e}_z$, and the lattice vectors are $\mathbf{R} = l\,a\,\mathbf{e}_z$, where a is the periodicity in the direction of the WG axis and l an arbitrary integer. The Bloch theorem Eq. 1.19 can be simplified for the PC-WG to:

$$\mathbf{H}(x,y,z) = \mathrm{e}^{-\mathrm{j}kz}\,\mathbf{u}(x,y,z) \text{ with } \mathbf{u}(x,y,z) = \mathbf{u}(x,y,z+a). \tag{1.20}$$

The first Brillouin zone also becomes somewhat simpler, it is just the wavevector-range $\mathbf{k} = -\frac{\pi}{a}\mathbf{e}_z \dots \frac{\pi}{a}\mathbf{e}_z$. Because of the mirror symmetry of the structure with respect to a plane $z = \text{const.}$, the eigenvalues for modes traveling in positive and negative z-direction are the same. This means it is sufficient to consider $k = 0 \dots \frac{\pi}{a}$. In analogy to the conventional strip-WG, k can as well be denoted as β, and the Bloch theorem Eq. 1.20 resembles the propagation equation of a mode in a conventional WG, Eq. 1.15. The band diagram of a PC-WG for TE-polarization is shown in Fig. 1.6. Two modes can be observed in the region of the bulk band gap. The silicon PC slab of height $h = 0.55\,a$ is surrounded by air, the hole radius is $r = 0.3\,a$, and the lattice constant a is scalable. The modes are separated into even and odd modes with respect

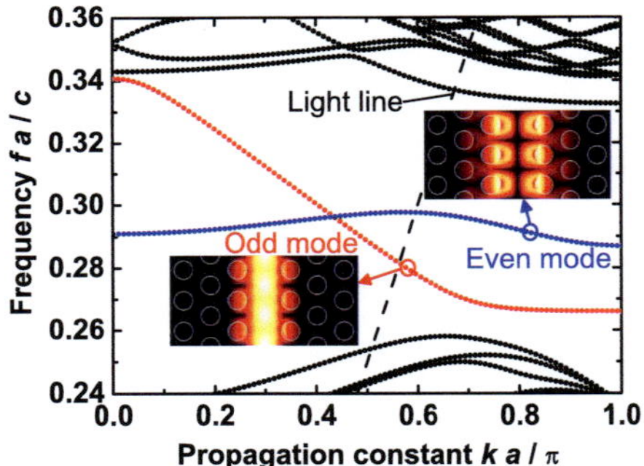

Fig. 1.6. Band diagram of a line defect PC-WG for the TE modes. Two defect modes appear inside the photonic band gap. The dominant electric field for a typical even and odd mode is shown as insets, and the field of the odd mode resembles the field of a fundamental strip-WG mode.

to a reflection in the yz-plane, which is the vertical plane that bisects the WG, see Fig. 1.3(b). Note that this classification into even and odd modes might be different from what is found in literature, where the classification is not consistent and even and odd might also refer to (quasi-) TE and TM modes, or even might correspond to our odd and odd to our even modes. The distribution of the dominant electric field component E_x for the even and odd mode is shown as insets, and the field of the odd mode resembles the field of a fundamental strip-WG mode. The light cone reduces to the light line, and modes below the light line are confined to the PC slab and have low losses, whereas modes above the light line become leaky. The defect modes of the PC-WG lie within the TE-bandgap of the bulk PC, but for low and high frequencies, the bulk bands appear. As there is no bandgap of the bulk PC for TM-polarization, we restrict ourselves to TE-operation.

For the modes of a line-defect PC-WG, the same orthogonality relation Eq. (1.17) holds as for the modes of conventional WGs. This is derived in Appendix A.3.

1.3   Group Velocity and Chromatic Dispersion

The phase fronts of a signal propagating in a strip-WG or a PC-WG travel with the phase velocity v_p, which is inversely proportional to the propagation constant β. The information in a signal is contained in the temporal changes of the signal amplitude and/or phase. In the simplest case, a logical 1 is represented by a short amplitude pulse, whereas for a logical 0 no pulse is present. The pulse centers however do not travel with the phase velocity v_p, but with the group velocity v_g, and the power flow is directly related to the group velocity. As pulses propagate along a WG, they might gradually change their shape, which is caused by the fact that different frequency components of the signal might travel with different group velocities. The change of the temporal width of a pulse during propagation can be quantified by the chromatic dispersion

C, and a WG or fiber with positive chromatic dispersion leads to pulse broadening, whereas negative chromatic dispersion can reverse the effect for a broadened pulse.

The phase velocity, the group velocity and the chromatic dispersion can be derived from the dependence of the propagation constant with frequency, $\beta(\omega)$. Around a carrier frequency $f_0 = \omega_0/(2\pi) = c/\lambda_0$, where λ_0 is the free-space wavelength, the propagation constant can be expanded into a Taylor series:

$$\beta(\omega) = \beta(\omega_0) + \beta'(\omega_0)(\omega - \omega_0) + \frac{\beta''(\omega_0)}{2}(\omega - \omega_0)^2 + \ldots \tag{1.21}$$

The first and second derivatives of β at ω_0 are denoted as $\beta'(\omega_0)$ and $\beta''(\omega_0)$, respectively. The relation for the *phase velocity* v_p, the *group velocity* v_g and the *chromatic dispersion* C as a function of frequency ω_0 is:

$$v_\mathrm{p} = \frac{\omega_0}{\beta(\omega_0)}, \tag{1.22}$$

$$v_\mathrm{g} = \frac{1}{\beta'(\omega_0)}, \tag{1.23}$$

$$C = \frac{-\omega_0^2}{2\pi c}\beta''(\omega_0). \tag{1.24}$$

The group delay t_g for a WG of length L is simply given by $t_\mathrm{g} = \frac{L}{v_\mathrm{g}}$. In analogy to the effective refractive index $n_\mathrm{eff} = c/v_\mathrm{p}$, an effective group index $n_\mathrm{g} = c/v_\mathrm{g}$ can be defined.

The *temporal pulse spread* $|\Delta t_\mathrm{g}|$ by chromatic dispersion can be expressed in an approximate formula, which is valid for long propagation distances L:

$$|\Delta t_\mathrm{g}| = CL\,|\Delta\lambda|\,. \tag{1.25}$$

The spectral width of the pulse in wavelength is $|\Delta\lambda|$, and the connection with frequency is $|\Delta\lambda| = \frac{c}{f^2}|\Delta f|$. The larger the bandwidth of a signal is, the larger is the pulse spreading by chromatic dispersion.

Modes in PC-WGs can exhibit regions with very low group velocity and very high chromatic dispersion [14]. From the band diagram of the PC-WG in Fig. 1.6, the group velocity and chromatic dispersion characteristics of the odd mode is calculated, Fig. 1.7. The frequency axis is normalized in the same way as in the band diagram, with a the period of the PC-WG and c the vacuum speed of light. At the points $ka/\pi = 0$ and $ka/\pi = 1$ in the band diagram, Fig. 1.6, the mode bands have a slope of 0, which implies a group velocity of zero. The slopes of 0 at these points results from the symmetry of the band diagram at $ka/\pi = 0$ and its periodicity in $k-$space. The odd mode hits the point $ka/\pi = 1$ near $fa/c = 0.265$ with a group velocity of 0, Fig. 1.7(a), whereas the chromatic dispersion is dramatically increased, Fig. 1.7(b). For increasing frequency, the group velocity increases and the chromatic dispersion decreases.

The geometry of a PC-WG can be adjusted such that unique group velocity or chromatic dispersion characteristics can be obtained, and in Section 2.1 we present details on how the design can be accomplished.

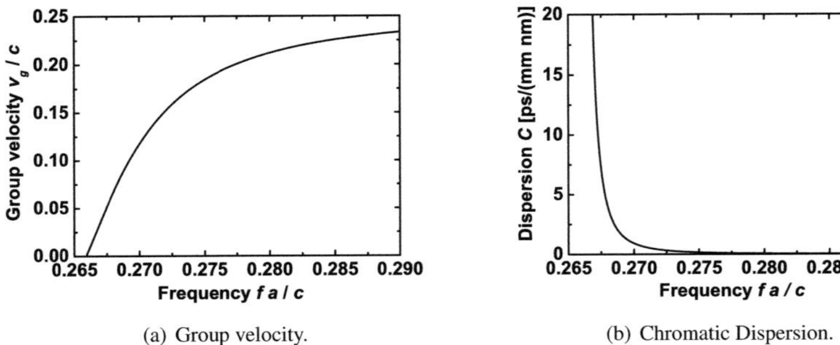

(a) Group velocity.   (b) Chromatic Dispersion.

Fig. 1.7. Group velocity and chromatic dispersion of a PC-WG. (a) The group velocity is normalized to the vacuum speed of light c. For (a) and (b), the frequency axis is normalized, where a is the PC period. For frequencies near the band edge $fa/c = 0.265$, the group velocity is very low and the chromatic dispersion very high.

The delaying and strongly dispersive properties of PCs can be used to build small functional devices like optical delay lines or dispersion compensators. In Section 3.1 of this work we present a tunable chromatic dispersion compensator, and in Section 3.2 a tunable optical delay line.

A second important effect of light propagation at low group velocity is that the interaction between light and matter is increased, which enhances nonlinear effects [15] and allows to decrease the size and to increase the efficiency of nonlinear devices like modulators, switches, wavelength converters or amplifiers. For an electro-optic device, the length to achieve a certain phase shift decreases proportionally with decreasing v_g, and the operational power for the smaller device also decreases. For an all-optical nonlinear device, the Kerr effect depends on the square of the electric field and increases inversely proportional with decreasing group velocity, and again at the same time the device length is decreased. Silicon is a centrosymmetric crystal, which means that there exist no $\chi^{(2)}$-effects, like the linear electro-optic effect. However, a $\chi^{(2)}$-effect in silicon can be induced by e. g. straining the crystal [16], or by infiltrating an electro-optic material into a silicon structure, like an electro-optic polymer [17] or a chalcogenide [18]. Furthermore, silicon exhibits $\chi^{(3)}$-effects, and stimulated Raman scattering [19], [20] and four-wave-mixing [21] have been demonstrated. In Section 3.3 we present a fast electro-optical modulator with low operation voltage.

Propagation losses, however, increase with decreasing group velocity, [15], [22], which might eventually limit the device performance. We especially address the issue of disorder-induced losses in PCs at low group velocity in Section 2.2 of this work.

1.4   Mode Gaps and Mini-Stop Bands in PC Waveguides

From the band diagram Fig. 1.6 it can be observed that inside the PBG there is a frequency range for which no mode can propagate in the WG, this is for frequencies below the cut-off of the odd mode near $fa/c = 0.265$. This region is called a mode gap [23], and it arises from

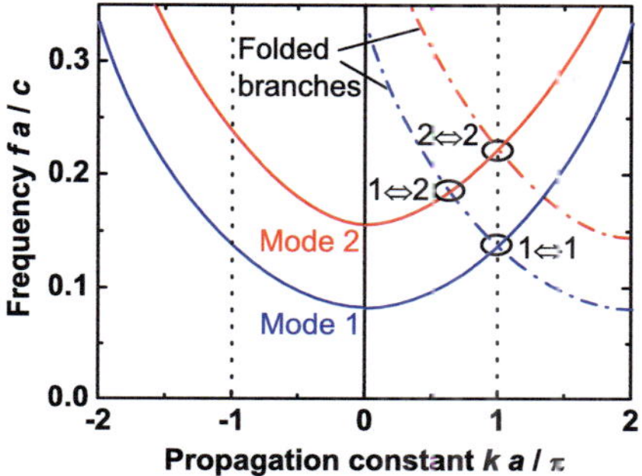

Fig. 1.8. Formation of mode gaps and mini-stop bands in a PC. If a perturbation of periodicity a is introduced to a WG, the mode dispersion curves are periodically repeated in k-space and folded to the first Brillouin zone. At the points where different bands intersect, $1 \Leftrightarrow 1$, $2 \Leftrightarrow 2$ and $1 \Leftrightarrow 2$, mode couplings can occur that lead to so-called mode gaps at the boundary of the first Brillouin zone, or to so-called mini-stop bands inside the first Brillouin zone.

an intrinsic coupling of counter-propagating modes. The same mechanism can also lead to the formation of so-called mini-stop bands at other k-points in the Brillouin zone [24], [25]. The mode gaps and mini-stop bands are the reason for the special group velocity and chromatic dispersion characteristics in a PC-WG [26] and are thus very useful; but also additional losses or reflections might occur in the case that the mode interactions are not intended. In a conventional strip-WG, the mode of lowest order has a certain cutoff-frequency, and for frequencies above the cutoff there is at least one mode that exists. For the strip-WG shown in Fig. 1.2, the cutoff frequency for the fundamental TE-mode is near 150 THz. The dispersion relation of a strip-WG can also be displayed in the same band diagram that is used for PC-WGs, the principle is shown in Fig. 1.8 for two propagating modes. If a periodic perturbation with periodicity a along the WG is introduced, the mode dispersion curves are periodically repeated in k-space with a periodicity of $\Delta k a/\pi = 2$ and can be folded to the first Brillouin zone $ka/\pi = -1\ldots1$. At points where bands intersect, mode couplings can occur. As stated in the previous section, the group velocity is $v_\mathrm{g} = \mathrm{d}\omega/\mathrm{d}\beta$, and at the different coupling points of mode 1 and mode 2, $1 \Leftrightarrow 1$, $2 \Leftrightarrow 2$ and $1 \Leftrightarrow 2$ in Fig. 1.8, the two modes have slopes of different sign and thus a group velocity of opposite direction. This means that the forward-propagating mode is coupled to the backward-propagating mode, and thus injected power will be reflected. The resulting modes of the WG with the periodic perturbation turned on will thus repel each other, and a so-called anti-crossing is formed, which leads to a frequency range without propagating modes. At the coupling points $1 \Leftrightarrow 1$ and $2 \Leftrightarrow 2$, where the branches of the same mode intersect, a mode gap emerges, which is basically nothing else than a Bragg reflection caused by the periodic perturbation. At the coupling point $1 \Leftrightarrow 2$, the branches of two different modes intersect, and a mini-stop band arises. The band diagram of a multi-moded PC-WG with a defect width that is doubled in comparison to the structure of Fig. 1.6 is displayed in Fig. 1.9. Inside the PBG $fa/c = 0.259\ldots0.341$, various even and odd WG modes exist. At the boundary of the first

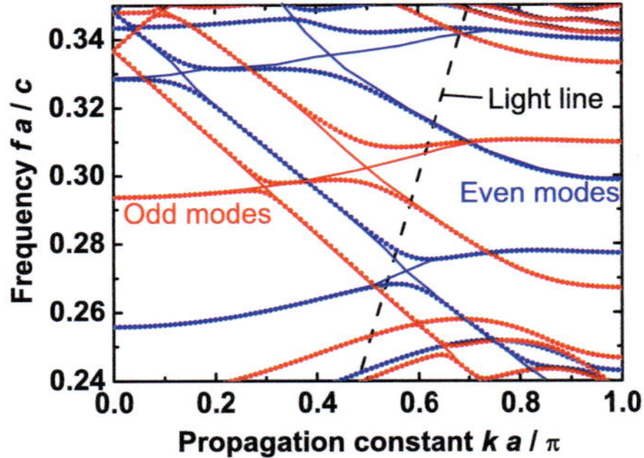

Fig. 1.9. Band diagram of a multi-moded PC-WG. At the boundary of the Brillouin zone $ka/\pi = 1$, mode gaps occur, and inside the Brillouin zone, mini-stop bands can be observed. The even and odd modes do not interact in the case of a vertical structure symmetry.

Brillouin zone $ka/\pi = 1$, a broad mode gap for the even and odd modes can be observed. Inside the Brillouin zone, several mini-stop bands can be seen, indicated by the solid lines that depict the mode curves before the periodic perturbation is introduced. However, even though a mode gap or a mini-stop band occurs for a certain mode, transmission might still be possible through another mode of the multi-moded PC-WG.

At the intersection points between the even and odd modes, no anticrossings are formed, which indicates that the even and odd modes do not interact. However, if the structure looses its symmetry with respect to a vertical plane along the WG-direction e. g. caused by fabrication-induced disorder, the even and odd modes can also couple, and unwanted transmission dips and reflection peaks might occur in the PC-WG [27]. Similarly, if a vertical structure asymmetry is induced, e. g. in the case of a SOI structure with glass as substrate material and air as cover material, or by hole walls that are not strictly vertical because of fabrication issues, the formerly uncoupled TE and TM modes can couple [28]. These two issues that might lead to unwanted mini-stop bands are more pronounced in slow-light PC-WGs because of the reduced bandwidth and the increased mode interaction time, and will be discussed in the Sections 2.2 and 4.1.

In a special PC-WG where additional holes are introduced in the WG center [13], [29], a highly negative chromatic dispersion can emerge. However, the fundamental mode shows an anti-crossing with the modes that can be used for excitation, and all power is reflected. A special coupling scheme is needed for such a PC-WG, as will be discussed in Section 2.1.

1.5   Excitation of Modes

The modes in a silicon PC-WG are strongly confined to a small geometrical area, which is due to high index-contrast in the SOI system. Typical heights of the silicon layer are 220 nm or 340 nm, which determines the vertical mode extension, and the horizontal mode extension is in the same order of magnitude. In an optical fiber, however, the index contrast is low, and the modal fields extend to a much larger core having a diameter of $9\,\mu$m. Thus, special measures have to be taken to efficiently couple light from a fiber to a silicon PC.

There are approaches for a direct coupling between a fiber and a PC-WG, e.g. through evanescent coupling of a thinned fiber in parallel to the PC-WG [30], [31]. However, it is more common to first couple from a fiber to a strip-WG, and then excite the PC mode with the mode of the strip WG. Efficient techniques for coupling fibers to silicon WGs have been well developed, e.g. through gratings [32], 2D tapers [33], [34] or 3D tapers [35]. High coupling efficiencies up to 80 % have been reached with these techniques.

For the coupling from the strip WG to the PC-WG, numerous methods have been proposed and demonstrated [36]. The two simplest yet effective approaches are the direct butt-coupling of the strip-WG to the PC-WG [34], and the introduction of a PC-taper in the first PC periods [37], [26]. These two concepts are schematically shown in Fig. 1.10. At the interface between the WG and the PC, the WG width or the position and radii of the PC holes can be modified to increase the coupling efficiency. A line-defect PC-WG that is generated by omitting a row of PC holes exhibits modes with spectral regions of moderate group velocity, and the mode fields are mainly confined to the WG center and resemble the mode fields of a strip-WG [38]. These modes can be efficiently excited by a butt-coupled strip-WG. However, in the spectral regions of low group velocity, the mode fields are significantly more extended to the regions of the PC holes, and a direct excitation with a strip-WG leads to a much worse efficiency. This is where the taper can lead to a significant improvement [39], [40]: In the first PC periods, the mode is gradually changed from a high to a low group velocity, and also its field distribution changes. In Section 2.1, we present details on how such a taper can be properly designed.

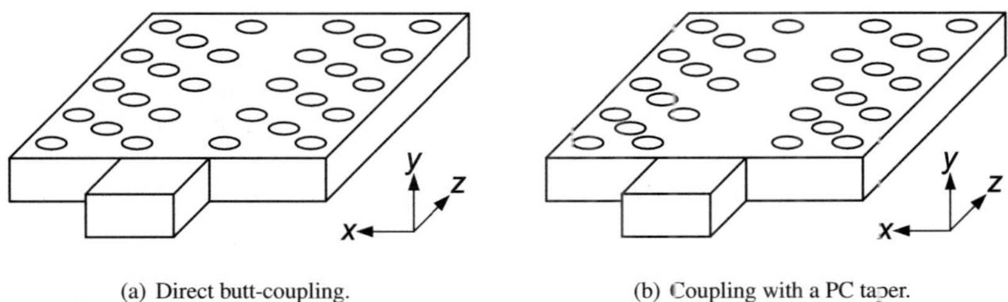

(a) Direct butt-coupling.   (b) Coupling with a PC taper.

Fig. 1.10. Coupling schemes from a strip WG to a PC-WG. The two simplest approaches are (a) butt-coupling and (b) the introduction of a PC taper, where the PC structure is gradually modified over a length of some PC periods.

1.6 Losses in Photonic Crystals

In the current stage of research, transmission losses in PCs represent the major limitation for integrated SOI-PC devices. The high index contrast of silicon towards air or glass leads to a strong light scattering at structure imperfections, and as the structure features have minimum sizes in the order of some ten nanometers, the accurate fabrication is a challenge. However, in the last years, great improvements in the disorder-induced losses of silicon strip WGs and PC-WGs have been achieved. The losses for a SOI strip WG of width $w = 500\,$nm and of height $h = 220\,$nm are 2.4$\,$dB$\,$/$\,$cm [41] and the losses for a membrane W1 PC-WG with slab height $h = 220\,$nm, hole radii $r = 240\,$nm and a lattice constant $a = 430\,$nm could be reduced to 4.1$\,$dB$\,$/$\,$cm [42]. Further improvements are still to be expected. For comparison, the losses of a standard SMF are $0.15 \times 10^{-5}\,$dB$\,$/$\,$cm. The losses in PC-WGs can be divided into intrinsic and extrinsic losses.

The *intrinsic losses* are the inherent losses of the WG, which are the linear and nonlinear material losses of silicon, and the leakage losses of the modes out of the slab if the condition of total internal reflection (TIR) is no longer met. The linear material losses of silicon are negligible [43]. The important nonlinear losses are two-photon absorption (TPA) [44] and free-carrier absorption (FCA) [45], and TPA can induce FCA [46]. For passive PC devices operated at low power levels, these nonlinear losses do not play a role either. The condition for TIR in the slab is met if the modes of the PC-WG stay below the light-lines of the substrate and the cover material [11]. In that case, there are no leakage losses of an ideal structure [47]. However, as soon as a mode crosses the light line, it can couple to radiative modes in the substrate or the cover and considerable losses can occur [48], [49]. A mode lying above the light line thus becomes a leaky mode, which is similar to a leaky mode in a fiber or a strip WG. The light line represents the dispersion of a wave that propagates in a homogeneous medium with refractive index $n_{\parallel}$:

$$k_{\parallel} = n_{\parallel}\,\frac{\omega}{c} \qquad \text{or} \qquad \omega_{\parallel} = \frac{kc}{n_{\parallel}}. \tag{1.26}$$

If the substrate and cover materials have different refractive indices, the larger refractive index needs to be considered as $n_{\parallel}$. This is the case as a larger refractive index leads to a light line that lies lower in the band diagram, and as soon as the lowest light line is exceeded, leakage losses arise. The higher the background index of a PC-WG is, the smaller is the low-loss frequency and wavevector range of the WG mode. Therefore it is desirable to choose a low background index, like e. g. air.

The *extrinsic losses* are the losses that are caused by non-ideal structures. They include fabrication-induced effects like insufficient hole depth [50], non-vertical sidewalls [28], [50], surface roughness [51], [52], disorder in the hole centers or hole radii [53], [54], and material inhomogeneities. In a PC-WG, the extrinsic losses can be separated in several contributions, see Fig. 1.11. We can distinguish between out-of-plane losses, in-plane losses and backscatter losses into the counterpropagating mode, and denote corresponding power loss factors α_{o}, α_{p} and α_{b}, respectively. At the interface between a strip-WG or a homogeneous material to a PC-WG, power can also be diffracted into space. If the PC-WG is operated inside the PBG and a sufficiently large number of holes is placed next to the WG, the in-plane losses α_{p} are small compared to α_{o} and α_{b} even in the presence of disorder or roughness, as the PBG of the bulk PC is relatively robust towards disorder and a large degree of disorder is needed to spoil the PBG [55], [56]. Currently, the losses in PC devices are mainly caused by imperfect fabrication processes. For line-defect PC-WGs, it is suggested that the transmission losses scale quadratically with the disorder variance σ [57], and that the losses are proportional to at least the first power of the group index [58], [59]. The knowledge of this fact is very important for operation at low group velocities, as it might introduce severe device limitations. In Section 2.2, we study the influence of radial disorder on the characteristics of slow-light PC-WGs.

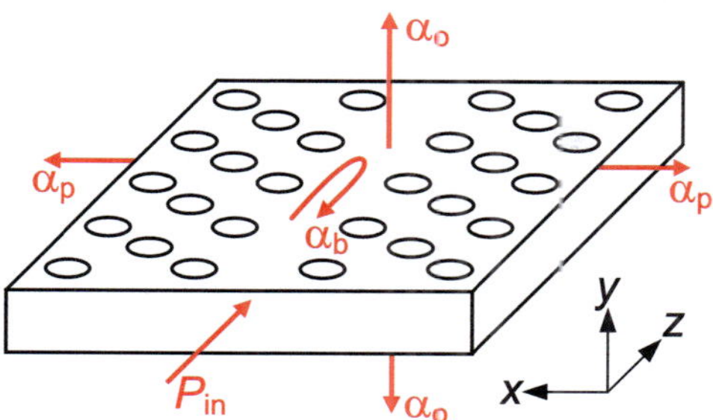

Fig. 1.11. Extrinsic losses in a PC-WG. The injected power P_{in} can be scattered out-of-plane with power loss factor α_o, in-plane with α_p, and back into the counterpropagating mode with α_b.

1.7 Tuning of Photonic Crystals

Various methods have been investigated and used to dynamically tune the properties of silicon photonic devices and especially PCs by an external control. In most cases, silicon is thermally tuned [60] or free carriers are injected into the silicon [61]; in fewer cases tunable materials like liquid crystals [62] or electro-optic polymers [17] are infiltrated into the silicon structure. Here, we give a brief overview on the thermal tuning, the free-carrier injection and the electro-optic effect.

1.7.1 Thermal Tuning

The change of the optical properties in silicon with changing temperature is caused by the linear expansion or contraction of the material and by the temperature coefficient of the refractive index n [63], where a change of the bandgap with temperature changes the number of carriers and thus the permittivity. The influence of the linear expansion is very small and can be neglected, and the temperature coefficient for an optical wavelength $\lambda = 1.53\,\mu$m and a temperature $T = 293$ K has been measured to

$$\frac{dn}{dT} = 1.80 \times 10^{-4}\,\mathrm{K}^{-1}.$$

(1.27)

The thermal tuning is rather slow with tuning times in the order of ms or hundreds of μs.

1.7.2 Free-Carrier Injection

Free carriers in a semiconductor change according to the classical Drude theory the real and imaginary parts of the dielectric permittivity. The following relations can be found that link the change in refractive index Δn and the change in absorption $\Delta \alpha$ with the change in electron and hole concentrations ΔN and ΔP, respectively [45]:

$$\Delta n = -\frac{e_0^2 \lambda^2}{8\pi^2 c^2 \varepsilon_0 n_{\text{Si}}} \left(\frac{\Delta N}{m_{\text{n}}} + \frac{\Delta P}{m_{\text{p}}} \right), \tag{1.28}$$

$$\Delta \alpha = \frac{e_0^3 \lambda^2}{4\pi^2 c^3 \varepsilon_0 n_{\text{Si}}} \left(\frac{\Delta N}{m_{\text{n}}^2 \mu_{\text{n}}} + \frac{\Delta P}{m_{\text{p}}^2 \mu_{\text{p}}} \right). \tag{1.29}$$

The electron charge is e_0, the effective masses and mobilities of the electron and holes are m_{n}, m_{p} and μ_{n}, μ_{p}, respectively, n_{Si} is the refractive index of silicon and λ the free-space wavelength of the light. Experimentally, the parameters of Eq. (1.28) were determined for silicon at a wavelength $\lambda = 1.55 \, \mu\text{m}$ and at room temperature, and the exponent of ΔP was empirically corrected to 0.8:

$$\Delta n = - \left(8.8 \times 10^{-22} \, \text{cm}^3 \, \Delta N + 8.5 \times 10^{-18} \, \text{cm}^{2.4} \, (\Delta P)^{0.8} \right). \tag{1.30}$$

The fastest tuning times with free carrier injection are in the order of some tens of ps. The basic speed limitation comes from the lifetime of the carriers, however the speed can be increased by extracting the free carriers from the WG zone with an electric field.

1.7.3   Electro-Optic Effect

The linear electro-optic or Pockels effect is a nonlinear $\chi^{(2)}$-effect, where an applied electric field $\mathbf{E} = (E_1, E_2, E_3)$ changes the optical refractive index [64]. In terms of the nonlinear polarization $\mathbf{P} = (P_1, P_2, P_3)$, the Pockels effect can be described as

$$P_{\text{i}}(\omega) = 2 \sum_{\text{j,k}} \chi_{\text{ijk}}^{(2)} (\omega : 0, \omega) \, E_{\text{k}}(0) \, E_{\text{j}}(\omega). \tag{1.31}$$

A polarization $P_{\text{i}}(\omega)$ at frequency ω is generated from the interaction of an electric field $E_{\text{k}}(0)$ at frequency zero and an optical field $E_{\text{j}}(\omega)$ at frequency ω in the nonlinear medium, and the interaction strength is quantified with the nonlinear susceptibility $\chi_{\text{ijk}}^{(2)}$.

The electric field $\mathbf{E}$ can be related to the electric displacement $\mathbf{D}$ by the 3×3 impermeability matrix η_{ij}, which is the inverse of the dielectric permeability matrix $\varepsilon_{r_{\text{ij}}}$:

$$E_{\text{i}} = \sum_{\text{j}} \eta_{\text{ij}} D_{\text{j}}. \tag{1.32}$$

As the dielectric permeability $\varepsilon_{r_{\text{ij}}}$ is real and symmetric for an electro-optic material, the same holds for the impermeability η_{ij}, and the number of coefficients can be reduced $\eta_{11} = \eta_1$, $\eta_{22} = \eta_2$, $\eta_{33} = \eta_3$, $\eta_{23} = \eta_{32} = \eta_4$, $\eta_{13} = \eta_{31} = \eta_5$, $\eta_{12} = \eta_{21} = \eta_6$. The change in the impermeability $\Delta \eta_{\text{i}} = \Delta \left(\frac{1}{n^2} \right)_{\text{i}}$ through the electro-optic effect can be expressed by the electrooptic coefficients r_{ij} and the applied electric field:

$$\Delta \eta_{\text{i}} = \Delta \left(\frac{1}{n^2} \right)_{\text{i}} = \sum_{\text{j}} r_{\text{ij}} E_{\text{j}}. \tag{1.33}$$

For small changes of the impermeability η, the change of the refractive index n can be approximated by:

$$\Delta n_i \approx -\frac{1}{2} n_i^3 r_{ij} E_j . \qquad (1.34)$$

For electrooptic polymers, the largest electrooptic coefficient is r_{33}, and the direction 3 corresponds to the direction along which the poling field was applied to align the electrooptic chromophores. Very high electro-optic coefficients up to $170\,\mathrm{pm}/\mathrm{V}$ have been demonstrated in a modulator device operated at $\lambda = 1.55\,\mu\mathrm{m}$ [65].

The electro-optic effect is intrinsically very fast, and the speed limitation comes rather from external effects, e. g. from the electrical feed circuitry.

Chapter 2

Slow-Light Waveguides

Over the past several years, new methods to dramatically slow down the propagation velocity of light pulses have been discovered and experimentally proven, which drew a lot of attention and lead to intense research activities in the new field of 'slow light'. In addition to the basic interest in the physics of light control, there are many new practical applications of slow light, which include an increased light-matter interaction leading to enhanced nonlinearities [15], the direct use of large delays for optical signal processing and storage or for metrology [66], [67], and the exploitation of a high chromatic dispersion resulting from a strongly varying group velocity [14].

Slow light can be achieved with a multitude of different approaches. In resonant atomic systems, extremely slow light can be achieved using electromagnetically induced transparency (EIT) [68], [69], [70] however only at low temperature. At room temperature, a similar effect to EIT is exploited called coherent population oscillation (CPO), where slow light could be shown in crystals [71], [72], doped fibers [73] or semiconductors [74], [75], [76]. Furthermore, slow light could be demonstrated by using stimulated Brillouin scattering (SBS) [77], [78] stimulated Raman scattering (SRS) [79], [80] or by wavelength conversion and dispersion [81]. So far, all mentioned slow light mechanisms rely on material properties. Moreover, also special material structuring can be used to generate slow-light effects. This way, light propagation at low group velocity was presented in Bragg fibers [82] or Bragg waveguides [83], in ring resonators or coupled-resonator optical waveguides (CROWs) [84], [85], [86], and in photonic crystal devices [60], [87], [88].

In this work, we especially study the potential of silicon slow-light photonic crystal waveguides for application in broadband optical data transmission systems. To this end, we first need to gain deeper insight into the properties of slow-light photonic crystal waveguides, and this insight will later be required to develop more complex slow-light devices and to interpret measurement results. This chapter is in particular devoted to two important aspects. First, in Section 2.1 we discuss several useful PC-WG geometries and develop design principles to obtain functional PC-WGs with broadband slow light or high chromatic dispersion characteristics, and we present methods for an efficient mode excitation. Second, in Section 2.2 we study the losses in PC-WGs caused by fabrication imperfections, which represent a major performance limitation of PC devices. In particular, we examine the influence of radial disorder on broadband slow-light PC-WGs and we suggest approaches for loss minimization.

2.1 Design

The primary attraction of PCs in the time when intense research on 2D and 3D PCs began was the ability to confine and guide light very efficiently in a small volume, and to be able to build very compact bends. Thus, the design was primarily aimed at maximizing the bandwidth and minimizing losses in PC-WGs [89] and in WG bends [90], [91], which leads to WGs with rather high group velocities or to multi-moded WGs.

Some years later, research activities arose that aim at utilizing and optimizing the strongly dispersive properties of PC-WGs, and to obtain a low group velocity or a high dispersion. Slow light propagation with a very low group velocity of 0.1% of the vacuum speed of light [88], [92] and a very high chromatic dispersion [14] have been demonstrated. For use in optical communication systems, the devices need to be broadband, and by adjusting the WG geometry, light propagation at a group velocity of $2\% - 3\%$ of the vacuum speed of light on a THz bandwidth has been proposed and realized [26], [93], [94], [95]. Also, in more complex systems of two coupled PC-WGs, broadband slow light and negative chromatic dispersion have been achieved [29], [96], [39].

In this Section, we study in detail how broadband PC-WGs with low group velocity or with high chromatic dispersion can be obtained. We present design procedures, which are then used to suggest a broadband slow-light WG with low losses, Section 2.2, to design a tunable dispersion compensator, Section 3.1 and a tunable delay line, Section 3.2, and to present a fast and compact electro-optic modulator, Section 3.3. Experimental results at optical frequencies and at microwave frequencies are given in Sections 4.1 and 4.2, and the verification of the used numerical methods is shown in Section 4.3.

This Section is structured as follows: In Subsection 2.1.1, three different PC-WGs geometries and their characteristics are presented, which are the standard line-defect WG, the slot WG and the center-hole WG. In Subsection 2.1.2, we start with general design principles and the influence of geometry changes on the bands, and continue in Subsection 2.1.3 with more details on how the group velocity and chromatic dispersion can be engineered and how a broadband behavior can be reached. In Subsection 2.1.4, the efficient excitation of the modes with low group velocity is discussed, and guidelines for an adiabatic taper that significantly enhances the coupling are given. For the center-hole WG, an efficient mode excitation with a butt-coupled standard WG cannot be achieved, and we suggest a better method using a counter-propagating coupling scheme.

2.1.1 Waveguide Geometries

In the SOI material system, the structure consists of a substrate out of glass (silicon dioxide) with $n_{\mathrm{sub}} = 1.44$, of the silicon device layer with $n_{\mathrm{slab}} = 3.48$, and of an air cover with $n_{\mathrm{cover}} = 1$. The glass substrate can be locally removed under the WG region to obtain a vertically symmetric membrane structure with $n_{\mathrm{sub}} = 1$, or the top of the silicon slab can be covered with glass, $n_{\mathrm{cover}} = 1.44$, which also fills the holes. The height of the silicon slab h is in most cases fixed to the available silicon slab height $h = 220\,\mathrm{nm}$ of standard SOI wafers, and in special cases a height of $h = 340\,\mathrm{nm}$ can be obtained.

Line-Defect Waveguide

The geometry of the line defect WG is schematically shown in Fig. 2.1(a). It is based on a standard W1 line defect WG, which means that one row of holes is removed to form the WG core [13]. This is the most commonly used PC-WG, and various geometrical parameters can be changed to design the WG properties, Fig. 2.1(b). In [97], the width (W_1) of the WG was

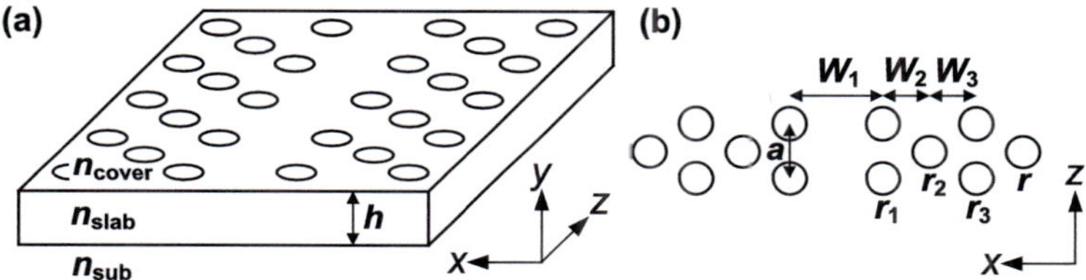

Fig. 2.1. Structure parameters of a line-defect PC-WG. The slab of width h and of refractive index n_{slab} is on top of a substrate material with n_{sub}, and is covered by a material with n_{cover}. The PC lattice constant is a, and the widths between the centers of the rows of holes next to the WG are W_1, W_2 and W_3. The hole radii of the rows of holes next to the WG are r_1, r_2 and r_3, and all other holes have radius r.

decreased to increase the bandwidth of the mode. In [26] the width (W_1) of the WG and the radii of all holes ($r_1 = r_2 = r_3 = r$) were changed, in [93] the radii of the first two rows of holes (r_1, r_2) were changed, and in [94] the radius of the first row of holes and of all other holes (r_1, $r_2 = r_3 = r$) were changed to obtain a broadband low group velocity. In [39], the width of the WG and between the first and second row of holes (W_1, W_2) were varied to achieve a high chromatic dispersion.

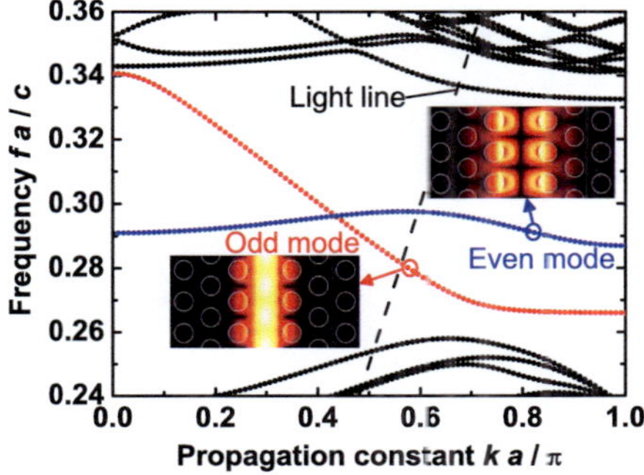

Fig. 2.2. Band diagram of a W1 line defect PC-WG. The dominant electric field for the even and odd mode is shown as insets. The silicon membrane has a height $h/a = 0.55$, the WG widths are $W_1 = \sqrt{3}\,a$, $W_2 = W_3 = \frac{1}{2}\sqrt{3}\,a$, and the hole radii are $r_1 = r_2 = r_3 = r = 0.3\,a$.

In our designs, we use more degrees of freedom to engineer the properties of the PC-WG, and depending on the requirements we choose to vary the width of the WG W_1, the width between the first and second and between the second and third row of holes, W_2 and W_3, respectively, or the radii r_1, r_2, r_3 and r of the first, second and third row of holes and the outer holes, respectively. Having more degrees of freedom also allows to better adapt to fabrication constraints, e. g. that holes should not be too small in diameter. The band diagram of a standard W1 silicon membrane PC-WG is displayed in Fig. 2.2 (repeated from Fig. 1.6). In the insets, the dominant field component $|E_x|$ is displayed for the odd and even modes. The separation into even and odd modes is with respect to a reflection in the yz-plane, which is the vertical plane that bisects the WG; please note that this classification might vary from what is found in literature. The structure parameters are $h/a = 0.55$, $W_1 = \sqrt{3}\,a$, $W_2 = W_3 = \frac{1}{2}\sqrt{3}\,a$, $r_1 = r_2 = r_3 = r = 0.3\,a$. The group velocity and chromatic dispersion characteristics are shown in Fig. 1.7. This WG geometry, the standard line-defect WG, can be adjusted to exhibit broadband low group velocity, and high positive or negative chromatic dispersion.

Slot Waveguide

The second WG geometry is the slot PC-WG. The schematic is shown in Fig. 2.3. In the center of a line-defect PC-WG, a gap of width W_{gap} is introduced, and the other geometrical parameters are the same than for the standard line-defect WG.

The slot of low refractive index causes a jump in the strength of the dominant electric field E_x, as the normal component of the electric displacement D_x is continuous across a dielectric boundary, and the mode field becomes very high in the gap region, especially for narrow gap widths. This idea has been proposed for a strip-WG in [98], and it can be very useful for nonlinear applications where the slot is filled with a nonlinear material [99]. Even for very small gap widths, the field stays strongly confined in the slot. The band diagram of such a structure is shown in Fig. 2.4, where the slab material is silicon and the cover and substrate material is glass. The structure parameters are $h/a = 0.54$, $W_{\mathrm{gap}}/a = 0.37$, $W_1 = 1.56\sqrt{3}\,a$, $W_2 = W_3 = \frac{1}{2}\sqrt{3}\,a$, $r_1 = r_2 = r_3 = r = 0.3\,a$. In the insets, again the dominant field component $|E_x|$ is displayed for the odd and even mode. It can indeed be observed that the odd mode is highly confined to the slot; this is however not the case for the even mode. The group velocity

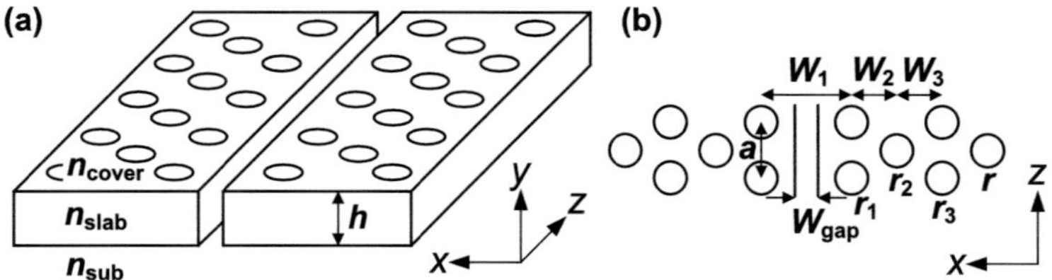

Fig. 2.3. Structure parameters of a slot PC-WG. The slab of width h and of refractive index n_{slab} is on top of a substrate material with n_{sub}, and is covered by a material with n_{cover}. A gap of width W_{gap} is introduced in the WG center. The PC lattice constant is a, and the widths between the centers of the rows of holes next to the WG are W_1, W_2 and W_3. The hole radii of the rows of holes next to the WG are r_1, r_2 and r_3, and all other holes have radius r.

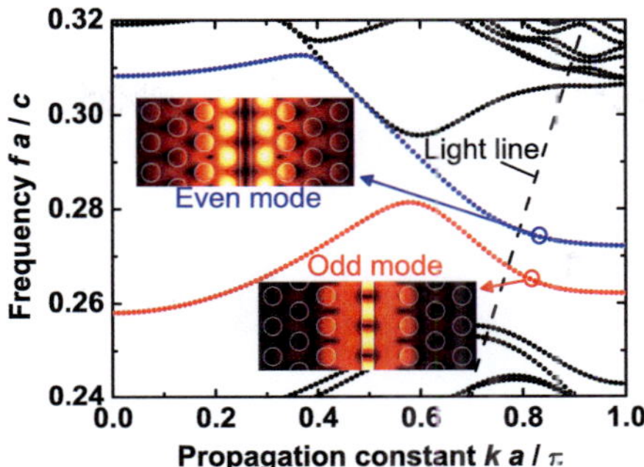

Fig. 2.4. Band diagram of a slot PC-WG. The dominant electric field for the even and odd mode is shown as insets. The silicon slab is surrounded by glass and has a height $h/a = 0.54$, the gap width is $W_{\mathrm{gap}}/a = 0.37$, and the WG widths are $W_1 = 1.56\sqrt{3}\,a$, $W_2 = W_3 = \frac{1}{2}\sqrt{3}\,a$. The hole radii are $r_1 = r_2 = r_3 = r = 0.3\,a$.

and chromatic dispersion characteristics of the odd mode show a similar behavior to the odd mode of the line-defect WG without slot, and the slot-WG geometry can be adjusted following the same principles than for the line-defect WG to exhibit broadband low group velocity or certain chromatic dispersion characteristics.

Center-Hole Waveguide

The third WG geometry we call the center-hole PC-WG. In the center of a line-defect PC-WG, an additional row of holes of radius r_c is introduced, Fig. 2.5 [13]. The specialty of this PC-WG is that it can exhibit modes with negative chromatic dispersion, and it has been proposed as part of an optical delay device [29], [96]. The WG can be also used for dispersion compensation,

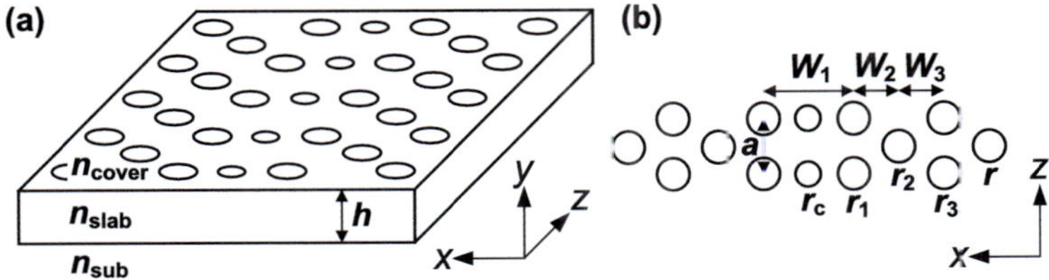

Fig. 2.5. Structure parameters of a center-hole PC-WG. The slab of width a and of refractive index n_{slab} is on top of a substrate material with n_{sub}, and is covered by a material with n_{cover}. A hole of radius r_c is introduced in the WG center. The PC lattice constant is a, and the widths between the centers of the rows of holes next to the WG are W_1, W_2 and W_3. The hole radii of the rows of holes next to the WG are r_1, r_2 and r_3, and all other holes have radius r.

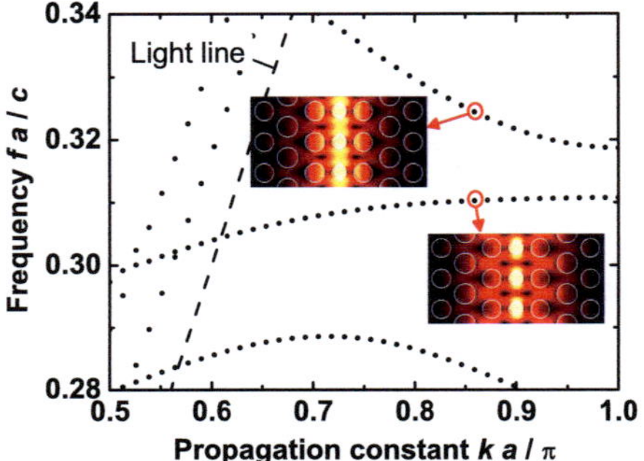

Fig. 2.6. Band diagram of a center-hole PC-WG. The dominant electric field for two odd modes is shown as insets. The slab with $n_{\mathrm{slab}} = 3.04$ is surrounded by air and has a height $h/a = 0.6$, the other parameters are $W_1 = 0.9\sqrt{3}\,a$, $W_2 = \frac{1}{2}\sqrt{3}\,a$, $W_3 = 0.87\,\frac{1}{2}\sqrt{3}\,a$, $r_{\mathrm{c}} = 0.23\,a$, $r_1 = r_3 = r = 0.3\,a$ and $r_2 = 0.329\,a$.

and a very high negative chromatic dispersion can be achieved. The band diagram of a center-hole WG is shown in Fig. 2.6, and only the odd modes are displayed. The membrane structure surrounded by air has a refractive index $n_{\mathrm{sub}} = 3.04$, and the structure parameters are $h/a = 0.6$, $W_1 = 0.9\sqrt{3}\,a$, $W_2 = \frac{1}{2}\sqrt{3}\,a$, $W_3 = 0.87\,\frac{1}{2}\sqrt{3}\,a$, $r_{\mathrm{c}} = 0.23\,a$, $r_1 = r_3 = r = 0.3\,a$ and $r_2 = 0.329\,a$. The mode with negative chromatic dispersion is the lower band near the normalized frequency $f\,a/c = 0.3$. The distribution of the dominant field component $|E_x|$ for the lower and the higher band are very similar, and the field is high in the region of the center hole, again because of the continuity in the D_x-component leading to an enhancement in the E_x-component in the material with low refractive index. The mode can be designed such that a broadband negative chromatic dispersion can be obtained. However, we find that the mode cannot be easily excited by a standard line-defect PC-WG or a strip-WG, whereas this is possible for the higher order mode. This seems peculiar as the mode profiles of the lower and higher mode are very similar. This issue will be discussed in more detail in Subsection 2.1.4, and a special excitation scheme is suggested. Because of the complexity of the mode excitation and of a high sensitivity of the mode behavior on structure changes, the center-hole WG will be discussed only briefly in this work.

2.1.2   Design Principles

In order to design a PC-WG with certain dispersive characteristics, it must first be studied how a change of the structural parameters affects the shape of the mode in the band diagram. In the following, the geometrical parameters W_1, W_2, W_3 and r_1, r_2, r_3 of a standard W1-line defect WG will be varied, and the change in the band diagram is observed, Fig. 2.7(a)-(d). The changes are very similar for the line-defect WG and the gap WG; thus we mainly concentrate on the line-defect WG here. The center-hole WG will not be treated in detail. The parameters that change the WG geometry near the core have a larger effect on the mode characteristics

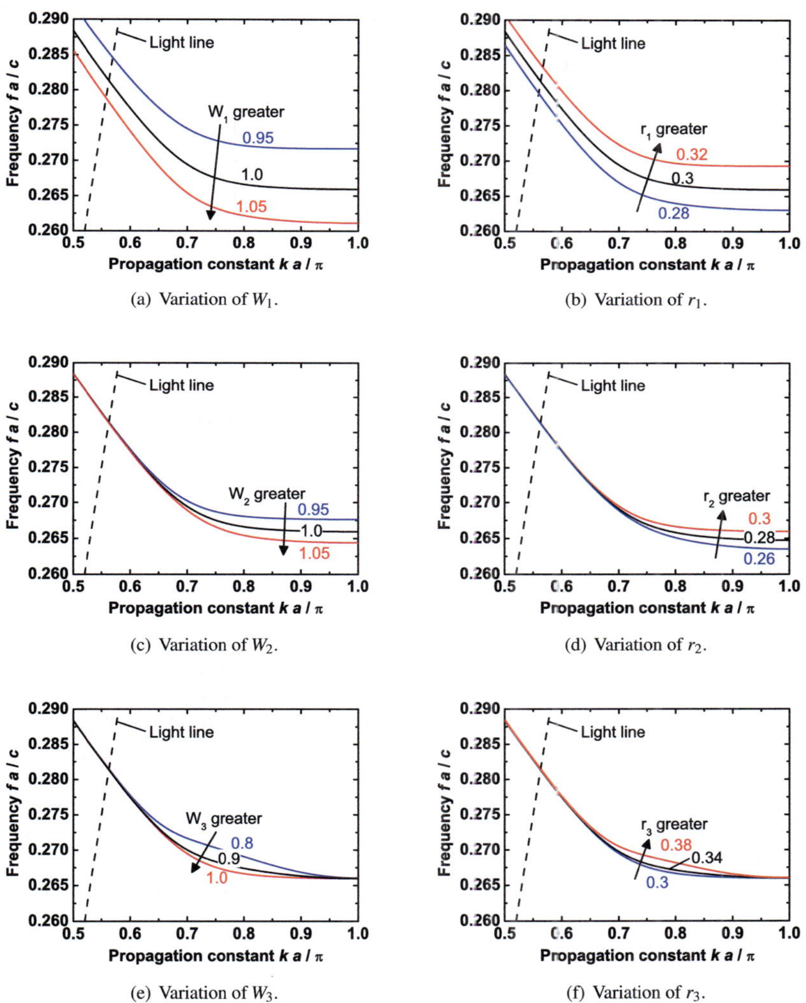

(a) Variation of W_1.　　　(b) Variation of r_1.

(c) Variation of W_2.　　　(d) Variation of r_2.

(e) Variation of W_3.　　　(f) Variation of r_3.

Fig. 2.7. Effect of structure parameter variation on the mode band of a W1-WG. The geometrical parameters W_1, W_2, W_3 and r_1, r_2 and r_3 are varied separately.

than the geometrical changes farther away from the core, which is clear as the mode is mainly confined to the core and the field strength decreases significantly in the region between the first and second row of holes, and even more between the second and third row of holes. However, as the group velocity decreases, which happens as the band approaches the edge of the first Brillouin zone $ka/\pi = 1$, the fields extend more into the patterned PC region. The geometry changes farther away from the core affect the band shape most strongly in the region of low group velocity.

A change in the width between holes W_i, $i = 1...3$, or in the corresponding hole radius r_i has a similar effect on the band deformation, which means that a particular characteristic can be reached by changing W_i, or r_i, or a combination of both. This allows to also take fabrication constraints into account, e. g. if a certain hole radius would become too small to be fabricated accurately, the design can be improved by increasing the hole radius and at the same time increasing the corresponding width between holes.

For the band diagrams shown in Fig. 2.7(a)-(d) and in Figs. 2.2-2.6, the mode band of interest below the light line is monotonously decreasing or increasing with propagation constant. For certain structure parameters, however, the band might become non-monotonous, which results in spectral ranges with two possible propagation constants at one frequency. This leads to a mode splitting into two branches at the k-point where the band slope is zero, as in the two regions the group velocity is of opposite sign (because of opposite signs of the band slope), where one of the mode branches cannot be easily excited. Thus, we try to avoid such bistable bands.

The highest refractive index of the background, n_{sub} or n_{cover}, determines the position of the light line, and decreasing the refractive index of the background moves up the light line in the band diagram, Eq. (1.26). A mode exhibits low intrinsic losses in the region below the light line, which in the band diagram is the wavevector-range between the intersection point of the mode with the light line and the edge of the first Brillouin zone $ka/\pi = 1$. To increase this range, a low background index can be chosen, and the membrane structure with $n_{\mathrm{cover}} = n_{\mathrm{sub}} = 1$ would be the best choice. However, a free-standing membrane structure might be more fragile and additional fabrication steps are needed to remove the substrate. For certain applications, even a higher background index must be accepted, e. g. if the WG is covered with a nonlinear polymer the background index increases to $n_{\mathrm{poly}} \approx 1.6$.

Another possibility to increase the range below the light line is to move the mode down in the band diagram, which can be achieved by increasing the width W_1. However, if W_1 is made too large, the mode might overlap with bulk PC modes appearing at low frequencies, and the WG might also become multi-moded, which is not desirable for applications where the group velocity or chromatic dispersion are the important characteristics. The disorder-induced losses of a PC-WG also decrease in general with increasing W_1. This can be intuitively understood, as for a high WG width, the PC holes having fabrication-induced disorder or roughness are geometrically farther away from the WG core region with high mode fields. The extrinsic loss mechanisms will be discussed in more detail in Section 2.2.

A third possibility to move the mode down in the band diagram is to choose a larger silicon slab height h; however the standard silicon wafer height is $h = 220\,\mathrm{nm}$ and other heights are rarely available.

The period a of the PC acts in the design process as a scaling parameter which shifts the operating wavelength to the desired value near $\lambda = 1.55\,\mu\mathrm{m}$. This is the case because of the scaling law of Maxwell's equations Eq. (1.12); however for a constant slab height, the wavelength scales only approximately linearly with the lattice constant.

After the third row of holes next to the WG core, a bulk PC follows. The hole radius r is chosen such that the PBG of the bulk PC is sufficiently large and overlaps with the frequency range of the WG mode; practical values are in the range $r/a = 0.25...0.35$. By increasing the value of r, the PBG is enlarged and shifted up in frequency. The PBG is in most cases much larger than the operating range of the mode.

2.1.3 Group Velocity and Dispersion Engineering

Broadband Slow Light

The term broadband slow light refers to a WG mode with low group velocity in a significantly large spectral range. More precisely, we mean a low group velocity or a high group index with only small variations inside the spectral range, which results in a low chromatic dispersion. In Fig. 2.8, the band diagram, the group velocity and the chromatic dispersion for a typical broadband slow light PC-WG are shown. In the band diagram Fig. 2.8(a), the mode has a constant low group velocity in the region where the band has a small slope and varies approximately linearly with propagation constant. At the low frequency end of the band where it hits the edge of the first Brillouin zone, the group velocity goes to zero and the chromatic dispersion to a very high value, Fig. 2.8(b). In the broadband slow light region, the group velocity has a value of a few percent of the vacuum speed of light, and the chromatic dispersion reaches a value near zero for a certain frequency. For higher frequencies, a local maximum of the chromatic dispersion is found, and the group velocity increases.

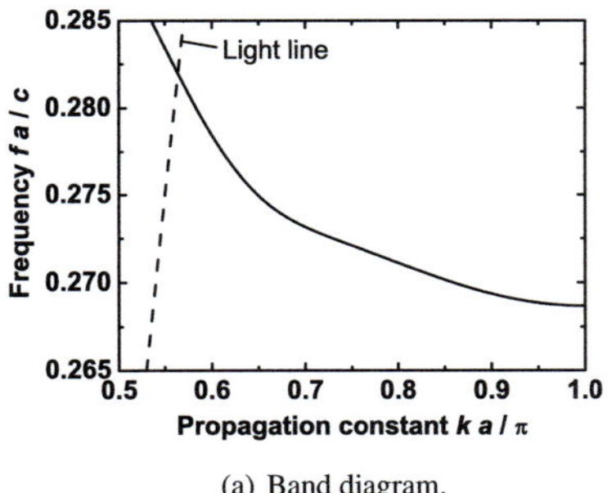

(a) Band diagram.

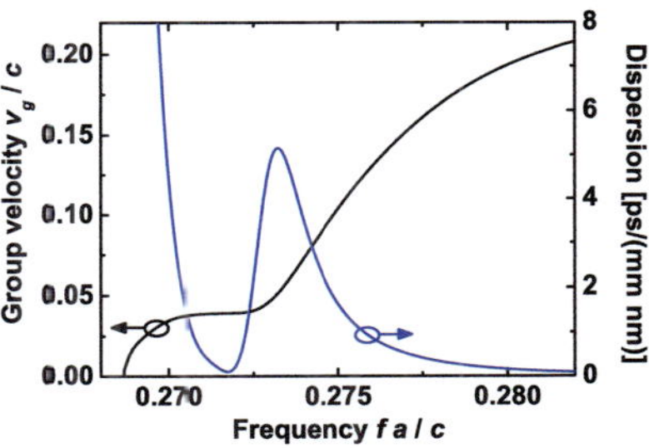

(b) Group velocity and chromatic dispersion.

Fig. 2.8. Broadband slow light PC-WG. (a) In the region where the band varies approximately linearly and has a small slope, the group velocity is constant with a low value. (b) The group velocity has a value of 3.9% of the vacuum speed of light in a broad 1.9 THz frequency range, and the chromatic dispersion is low. The structure parameters are given in the text.

The bandwidth of the slow-light region depends on the value of the group velocity; for lower group velocity the bandwidth is decreased. This can be explained as follows: The available range of the propagation constant $\Delta\beta$ for a mode in the band diagram below the light line is fixed, and for a decrease of the group velocity $v_g = \frac{d\omega}{d\beta} \approx \frac{\Delta\omega}{\Delta\beta}$ the available bandwidth $\Delta\omega$ also reduces.

In our design process of a broadband slow light WG, we start from a W1-WG in a silicon membrane of height 220 nm with hole radii $r/a = 0.3$. First, we adjust the width W_1 and the radius r_1, which have most effect on the band, while the width W_3 is already reduced to obtain a more linear characteristic of the band, Fig. 2.7(e), to decrease chromatic dispersion. The radius r_1 is reduced to obtain a higher group velocity and to improve the linear characteristic, and W_1 is slightly decreased such that the mode is shifted up in the band diagram and does not overlap with the bulk bands. Also, r_2 is reduced to further increase the group velocity. To have the hole radii r_1, r_2 not too small, the width W_2 is also decreased. The characteristic shown in Fig. 2.8 was obtained for the following geometric values: $W_1 = 0.95\sqrt{3}\,a$, $W_2 = 0.9\frac{1}{2}\sqrt{3}\,a$, $W_3 = 0.85\frac{1}{2}\sqrt{3}\,a$, $r_1/a = 0.25$, $r_2/a = 0.27$, $r_3/a = r/a = 0.3$. The group velocity was decreased to 3.9 % of the vacuum speed of light with a tolerance of $\pm 10\%$ in a 1.9 THz bandwidth. For the given design, the radius r_2 can be used to adjust the value of the group velocity, and a decrease in r_2 increases the group velocity. For a changed r_2-value, the readjustment of r_1 can again minimize the chromatic dispersion. The presented design is one of the possibilities to obtain broadband slow light, similar characteristics can also be obtained with other parameter sets.

Negative Chromatic Dispersion

The chromatic dispersion for the broadband slow light WG reaches a minimum value close to zero. To obtain a negative chromatic dispersion, the mode band needs to become more concave. The typical band diagram, group velocity and chromatic dispersion of such a PC-WG is shown in Fig. 2.9. At the low frequency end of the band, the chromatic dispersion starts at a high positive value, and then decreases to a minimum with a negative chromatic dispersion value. The chromatic dispersion characteristics is adjusted such that it varies nearly linearly in a frequency range below the minimum position, and also in the frequency range above the minimum position where chromatic dispersion rises again. This behavior is used for the dispersion compensator suggested in Section 3.1. The characteristic can be obtained

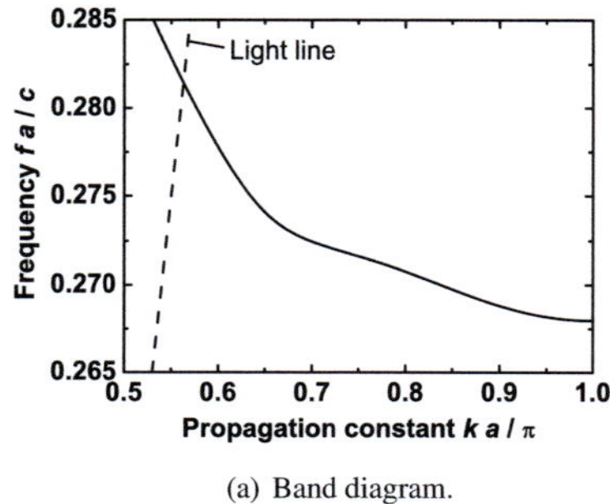

(a) Band diagram.

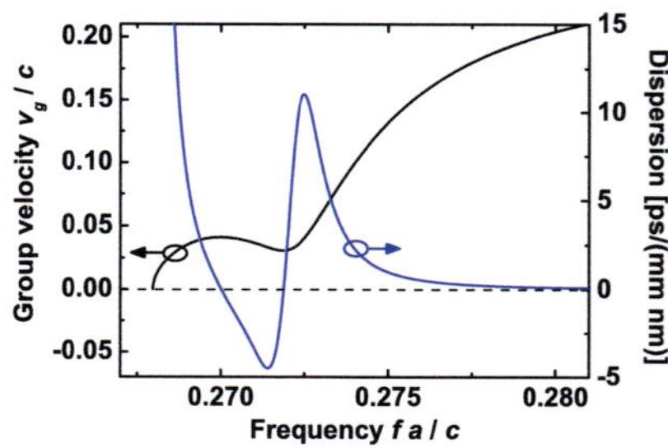

(b) Group velocity and chromatic dispersion.

Fig. 2.9. PC-WG with negative chromatic dispersion. (a) The band shape is slightly concave in the region of interest. (b) The chromatic dispersion has two nearly linear slopes in the frequency range before and after the minimum, and the minimum dispersion value is $-4.4\,\mathrm{ps}/(\mathrm{mm\,nm})$. The structure parameters are mentioned in the text.

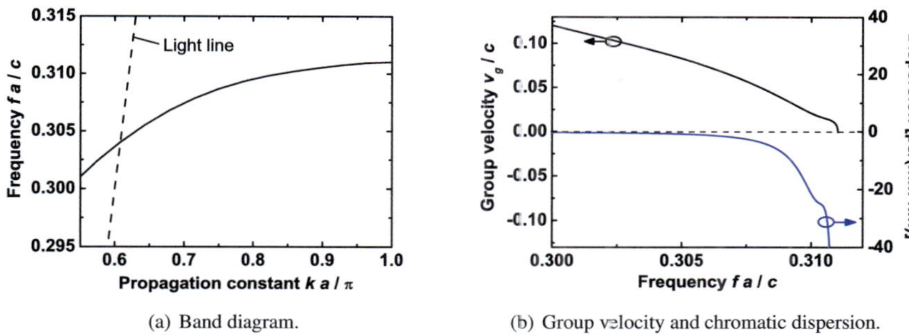

(a) Band diagram.

(b) Group velocity and chromatic dispersion.

Fig. 2.10. Center-hole PC-WG with negative chromatic dispersion. (a) The band is monotonously increasing with increasing propagation constant. (b) The decreasing group velocity characteristics with frequency leads to a negative chromatic dispersion in the full spectral range.

by decreasing W_3 and/or increasing r_3 starting from the design for broadband slow light. In addition, W_2 and r_2 have also been increased. The geometrical parameters are $W_1 = 0.95\sqrt{3}\,a$, $W_2 = \frac{1}{2}\sqrt{3}\,a$, $W_3 = 0.9\frac{1}{2}\sqrt{3}\,a$, $r_1/a = 0.25$, $r_2/a = 0.32$, $r_3/a = 0.34$, $r/a = 0.3$. The minimum dispersion is $-4.5\,\mathrm{ps}/(\mathrm{mm\,nm})$, and the maximum dispersion in the local maximum at higher frequencies is $11\,\mathrm{ps}/(\mathrm{mm\,nm})$.

Starting from a given design, the minimum dispersion values can be decreased and the maximum dispersion value increased by decreasing W_2 and increasing W_3 simultaneously; the group velocity is reduced at the same time. If the chromatic dispersion then deviates from the linear characteristics, adjusting r_1 can again lead to a linear behavior.

Another possibility to obtain a negative chromatic dispersion is by using the center-hole WG. The band diagram, the group velocity and the chromatic dispersion is shown in Fig. 2.10. The structure parameters are the same than for the center-hole WG presented in Subsection 2.1.1. The chromatic dispersion is negative in the whole region below the light line, which is an advantage over the previously presented line-defect WG with negative chromatic dispersion. However, the mode excitation is more involved and the WG design is more sensitive to parameter changes. For the center-hole WG, we will not present further details on the design procedure here.

2.1.4 Efficient Mode Excitation

Adiabatic Tapers

The fundamental mode of a standard W1 PC-WG has spectral regions of moderate group velocity with a field distribution similar to that of a strip-WG with a high field confinement to the WG core. The mode excitation in these spectral regions can be efficiently achieved by a butt-coupled strip WG. As the mode reaches the edge of the first Brillouin zone, the group velocity drops to zero in a narrow spectral range, and the coupling efficiency degrades considerably. At low group velocity, the fields inside the PC-WG extend more into the patterned PC region. For an efficient mode excitation at low group velocity, a coupling taper can significantly improve the coupling and reduce the reflection. With an adiabatic PC taper we mean a section of PC-WG where one or more geometrical parameters are gradually changed over some PC periods along propagation direction, Fig. 1.10(b), and the geometry change should be slow such that the mode can easily adapt to the changing structure.

The concept of such an adiabatic taper has been introduced in [37]. In a properly designed taper, the local mode in the starting section has a high group velocity and a field distribution similar to a strip-WG, which allows for an efficient coupling to a butt-coupled strip WG. Along the taper, the mode is gradually slowed down and changes its field distribution accordingly. In the band diagram Fig. 2.11(a), the local bands at the start and at the end of the taper and at intermediate points are displayed, and the field distribution at the taper start and taper end is shown as insets. The corresponding group velocity curves of the bands are plotted in Fig. 2.11(b). The band at the taper end is the slow-light mode into which the light should be coupled, and the dashed red line represents the operating center frequency of the final mode.

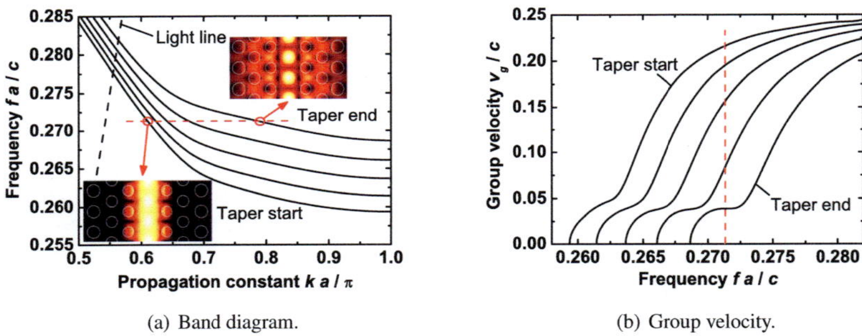

(a) Band diagram.   (b) Group velocity.

Fig. 2.11. Local modes of a PC taper. (a) Band diagram and (b) group velocity characteristics at several positions along the taper. The band at the taper end represents the mode into which the power will be coupled, and the taper start is butt-coupled to the strip WG. The group velocity is gradually reduced and the field profile gradually changed from taper start to taper end. The dashed red line represents the operating frequency.

Some points need to be considered when designing an efficient taper: At all positions along the taper, the local mode must be propagating and should be monotone avoiding a bistable behavior. Especially if more than one structure parameter is changed along the taper, it can happen that the bands at the taper start and taper end fulfill the requirement, but at some position inside the taper the mode does not propagate any more at the operating frequency or the mode becomes bistable. Also, the group velocity should continuously decrease along the taper, and not at any position along the taper be much lower than at the taper end. The modes along the taper can in some taper regions lie above the light line, which will increase the losses; but if the taper length is short this will not result in a severe degradation. In general, a longer and thus more gradual taper will increase transmission and decrease reflection; but in many cases a taper with a length of around 10 PC periods shows already good performance.

It has also been proposed that the coupling efficiency of a butt-coupled strip-WG to a slow-light PC-WG can be increased by modifying the PC termination [100]. We compare the transmission of direct butt-coupling with different termination and of excitation with a taper in Fig. 2.12(a). The operation frequency is around 193.5 THz, where the group velocity is 3.9 % of the vacuum speed of light. It can be clearly seen that the coupling taper has a significantly better performance than the best butt-coupled structure. In Fig. 2.12(b), the transmission and reflection of the structure with taper is displayed, and the transmission is better than -1 dB and the reflection lower than -10 dB in the frequency range of interest. In the simulations, the structure starts and ends with a strip-WG with a total of 30 PC periods in between, where in the case with tapers the length is 7 periods each for the starting and ending tapers. The slow-light WG structure is the broadband slow-light WG of Subsection 2.1.3 with $W_1 = 0.95\sqrt{3}\,a$, $W_2 = 0.9\frac{1}{2}\sqrt{3}\,a$, $W_3 = 0.85\frac{1}{2}\sqrt{3}\,a$, $r_1/a = 0.25$, $r_2/a = 0.27$, $r_3/a = r/a = 0.3$. At the taper start, W_1 and W_3 have been increased to $W_1 = 1.05\sqrt{3}\,a$ and $W_3 = 1.0\frac{1}{2}\sqrt{3}\,a$.

For all line-defect WGs and gap WGs presented in this work, we use an adiabatic coupling taper that is designed according to the principles mentioned in this section.

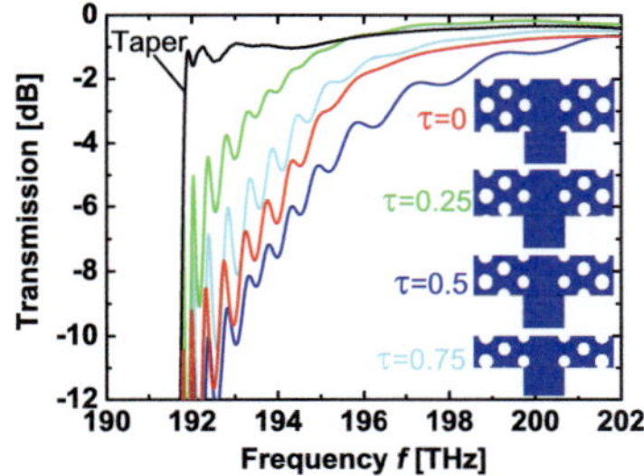

(a) Transmission with coupling taper compared to butt-coupling.

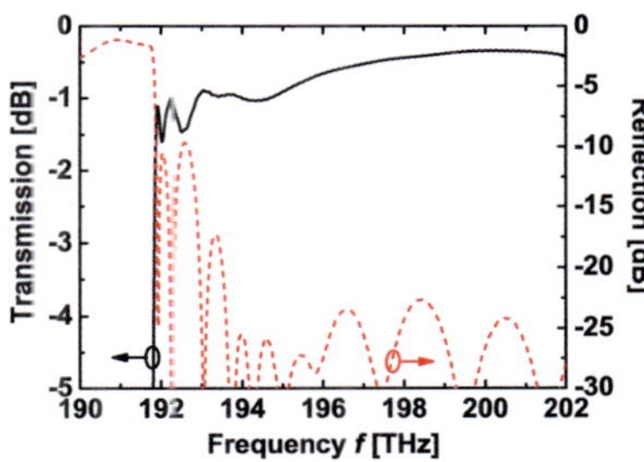

(b) Transmission and reflection with coupling taper.

Fig. 2.12. Coupling taper for efficient slow-light excitation. (a) The transmission with taper is higher than with the best butt-coupling structure (termination $\tau = 0.25$). (b) With taper, the transmission is better than -1 dB and reflection below -10 dB in the whole frequency range of interest. At 193.5 THz, the group velocity is 3.9 % of the vacuum speed of light..

Counter-Propagating Mode Coupling

In contrast to a conventional strip-WG, a PC-WG can exhibit modes where the phase velocity and the group velocity point in opposite direction. Two counter-propagating modes, i.e. two modes with opposite group velocity, might exist at the same frequency, and if they are coupled, a mini-stop band emerges. This issue was already introduced in Section 1.4. The mode of the center-hole WG with negative chromatic dispersion has the intrinsic property that if coupled to a strip-WG or a standard line-defect PC-WG, the group velocities of the center-hole WG mode and the coupling WG mode are in opposite direction. This can be observed from the band diagram Fig. 2.13.

The red line represents the mode of the center-hole WG, and the black line the mode of the line-defect WG. In the region below the light line, the two modes have an opposite slope of

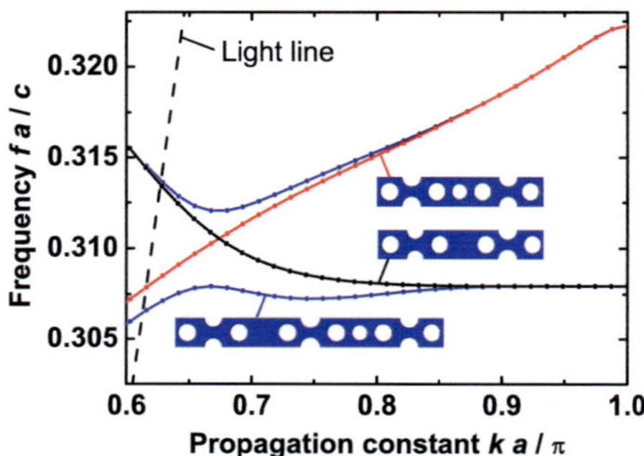

Fig. 2.13. Coupling of a center-hole WG to a line-defect WG. The band of the center-hole WG is displayed in red, and the band of the line-defect WG in black. If the center-hole WG and the line-defect WG are coupled in parallel with a spacing of three holes in between, a stop gap arises; the bands of the coupled WGs are displayed in blue.

the band and thus a group velocity of opposite direction. If the two WGs are parallel coupled e. g. with a spacing of three holes in between, a stop gap arises. This can be seen in the band diagram, where the supermodes of the two coupled WGs are represented by the blue lines. The emergence of the stop gap implies that if power is launched inside the gap, no mode can propagate in the coupled system. If in one of the WGs power is launched, it will be reflected back in the other WG. If the center-hole WG is excited with a butt-coupled strip-WG or line-defect WG, the power will be almost completely reflected or diffracted, which can be observed in simulations.

To excite the center-hole WG, we propose two methods, which both rely on parallel coupling of two WGs as shown in Fig. 2.13. Schematics of the two coupling methods are shown in Fig. 2.14, where the spacing between the WGs has been decreased to two rows of holes to increase the coupling. If power is injected in the line-defect WG inside the stop gap of the coupled system, the power in the line-defect WG decreases exponentially in propagation direction, while the power in the center-hole WG increases exponentially in the opposite direction, and vice versa. This property is directly utilized in the scheme Fig. 2.13(a), where the coupling section should be long enough so that all power is transferred from the line-defect WG to the center-hole WG. The 60°-bend in the line-defect WG helps to separate both WGs and is optimized for high transmission in the operating frequency range. In the scheme in Fig. 2.13(b), the coupling also occurs into the counter-propagating mode, however reflections take place at the positions where the defect WGs end, and the length of the coupling section can be adjusted such that all power is transferred from one to the other WG.

The coupling behavior can be modeled with a coupled-mode formalism [101], [102], which we adapt in Appendix A.5 to the counter-propagating case. The model can not only be used to explain the coupling phenomenon here, but can more generally describe the coupling of counter-propagating modes in PC-WGs as mentioned in Section 1.4.

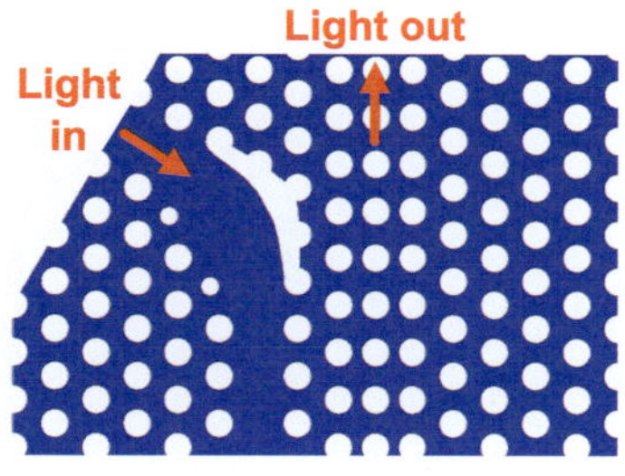

(a) Structure with long coupling section.

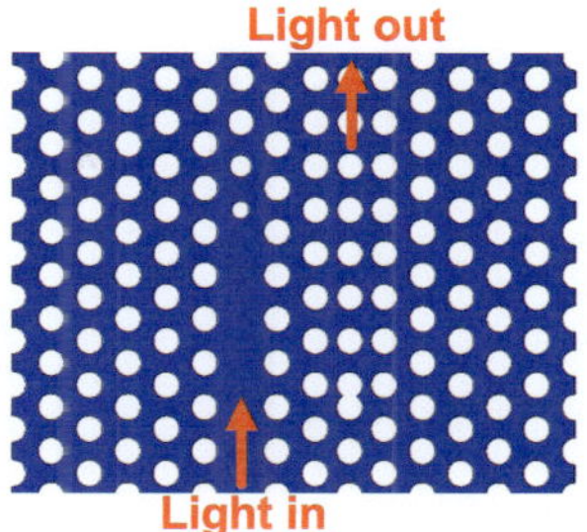

(b) Structure with short coupling section.

Fig. 2.14. Counter-propagating mode excitation of center-hole WG. Power is injected in the line-defect WG, and coupled to the center-hole WG with power flow in opposite direction. (a) For the long coupling section, the power decreases exponentially in the line-defect WG, while the power increases exponentially in opposite direction in the center-hole WG. (b) For the short coupling section, all impinging power is reflected at the positions where the individual WGs end. The length of the coupling section can be adjusted such that all power is transferred from one to the other WG.

2.1.5 Conclusion

We develop detailed design procedures to engineer the group velocity and chromatic dispersion characteristics of PC-WGs. Three different PC-WG geometries are studied, which are the line-defect WG, the slot WG, and the center-hole WG. Broadband slow light with a group velocity of 3.9 % of the vacuum speed of light in a 1.9 THz bandwidth and negative chromatic dispersion below $-4.5\,\mathrm{ps}\,/\,(\mathrm{mm\,nm})$ can be achieved by appropriately changing the structure parameters. We use a PC coupling taper to significantly improve the excitation of the slow-light PC mode with a strip-WG. The taper can decrease the loss per interface to below $-0.5\,\mathrm{dB}$, and reduce the reflection to a value smaller than $-10\,\mathrm{dB}$. Furthermore, a coupling concept is presented that allows exciting a mode of the center-hole WG with high negative dispersion by injecting power inside a mini-stop band.

2.2 Imperfections

Light propagation inside a PC-WG is theoretically lossless if the mode stays below the light line and if material losses are negligible. However, extrinsic losses caused by imperfect fabrication processes limit the performance of PC devices significantly. Disorder-induced losses in PC-WGs have considerably decreased over the last years, as much effort has been made to improve the SOI fabrication processes and to understand the loss mechanisms in PCs. The lowest losses reported in a silicon membrane W1-WG are $4.1\,\mathrm{dB/cm}$ [42], which is a value that can be tolerated for small integrated optical devices. The lowest losses in a PC-WG, however, occur at high group velocities, while disorder-induced losses considerably increase with decreasing group velocity. Based on a theoretical model supported by experimental data [59], the losses obey a quadratic law $\alpha \propto 1/v_g^2$ in the case of dominant backscattering losses occurring at low group velocities, and have a linear relation $\alpha \propto 1/v_g$ in the case of dominant out-of-plane losses at higher group velocities. Another experimental work [103] suggests a square root law $\alpha \propto 1/\sqrt{v_g}$ for dominant out-of-plane losses at group velocities above 5% of the vacuum speed of light. These different findings show that so far, the dependence of losses on group velocity is not well understood. Furthermore, methods to minimize losses for slow-light WGs have not been presented in detail.

The dependance of scattering losses on the disorder root mean square value σ is quadratic, $\alpha \propto \sigma^2$, which has been theoretically and experimentally well proven [57], [103].

In this section, we numerically investigate the effects of radial disorder on broadband slow-light PC-WGs. We seek to develop broadband slow-light WGs with minimum losses, which also requires to gain more insights into the effect of low group velocity on the disorder-induced losses. To this end, we design various broadband slow-light structures that all exhibit a group velocity of about 4% of the vacuum speed of light in a bandwidth of more than 1 THz. The disorder-induced losses of the designs are numerically determined and compared. Two numerical methods are used, which are the FIT and the GME methods, and the suitability and limitations of both methods are discussed. In order to minimize losses, we find that the PC-WG should be single-moded, otherwise disorder-induced mode coupling occurs that leads to increased losses. Furthermore, increasing the core width of a WG can decrease losses, as long as the WG is still single-moded. For a structure that is not vertically symmetric, like a SOI structure, intrinsic mode coupling between the previously orthogonal TE and TM modes occur, which can lead to mode gaps which considerably increase losses. A single-moded PC-WG with a vertically symmetric structure, like a membrane structure or a SOI structure with glass cladding, has thus lowest losses. The group velocity behavior of the broadband slow-light WG is only slightly changed by the presence of disorder, which is an important property. Furthermore, we examine the relation between disorder-induced losses and group velocity, but contrast to the above mentioned published results we find that simple relations like a quadratic, linear or square root law do not adequately reflect the behavior

This section is structured as follows: In Subsection 2.2.1, the parameters to characterize losses are specified and a model is introduced to determine transmission and reflection of a WG from these loss parameters. Subsection 2.2.2 discusses the two numerical methods FIT and GME to determine losses. In Subsection 2.2.3, the different broadband slow-light designs are presented, and in Subsection 2.2.4 the loss results are presented and discussed. Finally, the dependance of the losses on the group velocity characteristic is treated in Subsection 2.2.5.

2.2.1   Loss Parameters and Loss Model

The losses of a PC-WG can be quantified by loss parameters α, as already introduced in Section 1.6. The out-of-plane loss α_o results from the coupling of the WG mode to radiation modes that can propagate in the half-spaces below or above the silicon slab, which can be caused by mode operation above the light line or by the existence of structural disorder. The backscattering loss α_b is caused by the disorder-induced coupling of a forward propagating mode with the respective backwards propagating mode. The in-plane loss α_p quantifies the leakage of power out of the WG into the silicon slab plane, which can be caused by an insufficient number of row of holes next to the WG core, or by a missing or insufficient bulk bandgap in the presence of disorder. In our case, we expect the in-plane loss to be very small, and we model residual in-plane loss as part of the out-of-plane loss. If the PC-WG is multi-moded, disorder can also lead to coupling of the desired mode with other forward or backwards propagating modes. Similarly, a vertical structure asymmetry can couple the originally separable TE and TM modes, where the vertical asymmetry can be either intrinsic, as for an SOI structure, or again caused by disorder. The mode coupling to other guided modes also effectively increases losses, as the other modes exhibit undesirable group velocity characteristics, and as converted power is not coupled out to the connecting WGs. We will also model mode coupling effects as part of the out-of-plane loss. Thus, we characterize the losses by the two parameters α_o and α_b, and the total loss α_t is the sum of the two parameters, $\alpha_t = \alpha_o + \alpha_b$.

For a PC-WG of length L, the presence of losses decreases the transmitted power and increases the reflected power in comparison to the lossless structure. In all measurements and fully-numerical methods like FIT, the transmission and reflection characteristics are determined rather than the loss parameters. Thus, we need to introduce a loss model to be able to link the transmission and reflection with the loss parameters. The loss model is based on the coupled power equations of Marcuse [104], [105], and schematically shown in Fig. 2.15. The power traveling to the right P_r is diminished by the out-of-plane scattering α_o and by the backscattering α_b, while it is increased by the backscattering α_b from the mode traveling to the left. The analog behavior applies for the power traveling to the left, P_l. In the coupled power equations, the powers P_r and P_l stand for ensemble averaged values. In the derivation of the coupled power equations, perturbation theory is used, which is strictly valid only for weak coupling. However, even in the case of strong coupling, the equations can be used if the loss coefficients are fitted to measurement or simulation results. The differential equations that govern the power flow to the left $P_l(z)$ and to the right $P_r(z)$ as function of position can be formulated as:

$$\frac{dP_r(z)}{dz} = -(\alpha_b + \alpha_o)P_r + \alpha_b P_l, \qquad (2.1)$$

$$\frac{dP_l(z)}{dz} = (\alpha_b + \alpha_o)P_l - \alpha_b P_r. \qquad (2.2)$$

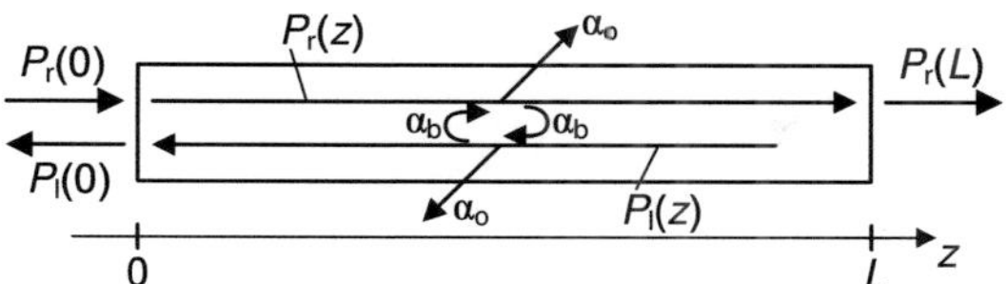

Fig. 2.15. Loss model for a PC-WG with out-of-plane and backscattering loss. The out-of-plane and backscattering loss parameters are a_o and α_b, respectively. The power traveling to the left and right is denoted as P_l and P_r, respectively, and the transmitted and backscattered power of a WG of length L is $P_r(L)$ and $P_l(0)$, respectively, for an injected power $P_r(0)$.

The differential equations Eq. (2.1) and Eq. (2.2) are solved using the boundary conditions $P_l(L) = 0$ (reflected power at $z = L$ is zero) and $P_r(0) = 1$ (injected power at $z = 0$ is unity), and the normalized transmission $P_r(L)$ and normalized reflection $P_l(0)$ are found (Appendix A.6):

$$P_l(0) = \frac{\alpha_b \tanh(\alpha L)}{\alpha_t \tanh(\alpha L) + \alpha}, \tag{2.3}$$

$$P_r(L) = \frac{\alpha_b}{\alpha}\sinh(\alpha L)P_l(0) + \cosh(\alpha L) - \frac{\alpha_t}{\alpha}\sinh(\alpha L), \tag{2.4}$$

$$\alpha_t = \alpha_b + \alpha_o, \quad \alpha = \sqrt{\alpha_t^2 - \alpha_b^2} = \sqrt{\alpha_o^2 + 2\alpha_b\alpha_o}. \tag{2.5}$$

The normalized powers $P_l(0)$ and $P_r(L)$ are in the range between 0 and 1, and the total power transmitted in positive z-direction is $P_{\text{tot}}(z) = P_r(z) - P_l(z)$.

The results in Eqs. (2.3)-(2.5) are hard to understand intuitively, and in order to gain more physical insight into the behavior the model Eqs. (2.1)-(2.2) is simplified:

$$\frac{dP_r(z)}{dz} = -(\alpha_b + \alpha_o)P_r, \tag{2.6}$$

$$\frac{dP_l(z)}{dz} = -\alpha_b P_r. \tag{2.7}$$

On the right hand sides of the equations, the terms involving the power traveling to the left are neglected, as the power traveling to the left should be much smaller than the power traveling to the right. The solution using the same boundary conditions as above is found to be (Appendix A.6):

$$P_l(0) = \frac{\alpha_b}{\alpha_t}\left(1 - e^{-\alpha_t L}\right), \tag{2.8}$$

$$P_r(L) = e^{-\alpha_t L}. \tag{2.9}$$

The transmission, Eq. (2.9), obeys the commonly used exponential relation. For a small product of total loss and length, $\alpha_t L \ll 1$, the backreflection Eq. (2.8) can be approximated by $P_l(0) = \alpha_b L$. For large WG lengths, the backreflection $P_l(0)$ approaches the upper limit α_b/α_t.

The lossy WG can also be represented as a linear two-port network where the first port is located at position $z = 0$ and the second port at $z = L$. The network can be described by its four scattering parameters S_{ij}, $i, j = 1, 2$, which relate the outgoing complex wave amplitudes b_j at port j with the incoming wave amplitudes a_i at port i by $b_j = \sum_i S_{ji} a_i$. The square of the absolute values of the scattering parameters correspond to the normalized powers of Eqs. (2.3)-(2.5) or Eqs. (2.8)-(2.9), $|S_{11}|^2 = P_1(0)$ and $|S_{21}|^2 = P_r(L)$.

2.2.2 Numerical Methods

For the numerical determination of disorder-induced losses, two methods are used, which are the GME and FIT methods, see also Appendix A.2. Here we discuss the suitability and reliability of the two methods for the simulation of losses.

The FIT is a fully-numerical method that solves Maxwell's equations at discrete times and in a discretized 3D space. In general, all possible structures can be simulated, and all linear effects that lead to losses are taken into account. Numerical errors are mainly caused by the limited structure discretization and a finite simulation time. We show in Subsection 4.3.2 that results from disorder simulations for both a single realization of a disordered structure and for ensemble averages agree very well with results from FDTD, which gives a high credibility to the FIT results. The main drawback of the method is, however, that the simulation time is very long, which limits the number of structures that can be simulated in an acceptable time.

On the other hand, the GME method is very fast, as it is a semianalytical method that uses a perturbative approach to determine losses. However, it is not an exact method and only considers coupling to radiative modes and to the counter-propagating mode, whereas mode coupling with other guided modes is not taken into account. Thus, the losses of multi-moded or vertically asymmetric WGs can not be reliably simulated with the GME method. Also for single-moded and vertically symmetric WGs, we expect the loss results from GME to be less accurate than those from FIT, as outlined in Subsection 4.3.2. But if properly used, the GME method allows a quick comparison of a large number of structures.

2.2.3 Broadband Slow-Light Structures

The disorder-induced losses of a PC-WG show a strong dependance on the group velocity of the mode, as already mentioned in the introduction of the section. Thus, to compare the losses of different slow-light structures, the group velocity characteristics need to be similar. If the structure of a PC-WG is optimized to exhibit low losses, e. g. using automatic algorithms, it will ultimately lead to designs with high group velocity. Here, we choose to fix the group velocity behavior and find different designs based on a line-defect WG that lead to a similar characteristics.

We show six different broadband designs that exhibit a group velocity of 3.7 % of the vacuum speed of light at an operating frequency of 193.4 THz or an operating wavelength of 1.55 μm. In addition, a standard W1-WG is chosen as reference. The broadband slow-light WGs are designed with the GME method according to the principles presented in Section 2.1. The lattice constant a for all designs is adjusted such that the operating frequency is at 193.4 THz. The structures with the corresponding parameters are summarized in Table 2.1, and structure schematics are shown in Fig. 2.16.

Parameter	Air_Nar	Air	Air_Broad	Bur	Bur_Thick	SOI	Ref
Refractive index							
n_{sub}	1	1	1	1.44	1.44	1.44	1
n_{cover}	1	1	1	1.44	1.44	1	1
Width							
W_1 (in $\sqrt{3}\,a$)	0.75	0.95	2	0.95	0.95	0.95	1
W_2 (in $\frac{1}{2}\sqrt{3}\,a$)	1	0.9	0.8	0.9	0.9	0.9	1
W_3 (in $\frac{1}{2}\sqrt{3}\,a$)	1	0.86	0.85	0.95	0.95	0.87	1
Radius							
r_1 (in a)	0.235	0.25	0.25	0.25	0.25	0.25	0.3
r_2 (in a)	0.235	0.27	0.33	0.27	0.27	0.27	0.3
r_3 (in a)	0.235	0.3	0.3	0.3	0.3	0.3	0.3
r (in a)	0.235	0.3	0.3	0.3	0.3	0.3	0.3
Height h (in nm)	220	220	220	220	340	220	220
Period a (in nm)	449	420	427	406	370	417	422

Table 2.1. Structure parameters for broadband slow light designs. All structures except for the 'Ref' structure are designed to have a group velocity of 3.7 % of the vacuum speed of light at 193.4 THz in a large bandwidth.

The first three designs 'Air_Nar', 'Air' and 'Air_Broad' are silicon *air* membrane structures, where the cladding and the substrate material is air, and the silicon slab height is 220 nm. 'Air_Nar' and 'Air_Broad' have a *narrow*er and *broad*er core width in comparison to the 'Air' design. The light line index for the membrane structures is $n_{\parallel} = 1$. The band diagrams of the three membrane structures are shown in Fig. 2.17(a)-(c). The main difference is the width of the WG core W_1, and the other structure parameters are adjusted such that the group velocity behavior is flat. The structure 'Air_Nar' has a narrow core of width $W_1 = 0.75\sqrt{3}\,a$ and equal hole radii for all holes; it is single-moded. For the structure 'Air', the core width has been increased to $W_1 = 0.95\sqrt{3}\,a$ to be as large as possible to be still single-moded. It can be observed that the band of the 'Air' structure is shifted down in normalized frequency in comparison to the 'Air_Nar' structure, which increases the operation range below the light line. The structure 'Air_Broad' has a broad core of width $W_1 = 2\sqrt{3}\,a$ and is multi-moded, in addition to the mode of interest another odd and three other even modes exist in the operation frequency range. The *reference* structure 'Ref' is a silicon membrane W1 PC-WGs, which is widely used in the literature and added for comparison.

The designs 'Bur' and 'Bur_Thick' are *buried* vertically symmetric structures where the substrate and cladding materials are glass, and also the holes are filled with glass. The 'Bur_Thick' design has a *thick*er slab width of 340 nm in comparison to 220 nm for the 'Bur' design, while all other stucture parameters are same. The light line index for the buried structures is $n_{\parallel} = 1.44$, which means that the operation range under the light line is reduced in comparison to the membrane structures. The band diagrams of the 'Bur' and 'Bur_Thick' structures are very similar to the 'Air' band diagram, Fig. 2.17(b), except for the lower light line of the buried structures. This is the reason why we do not display the band diagrams here.

The SOI design 'SOI' has a vertical asymmetry, the substrate material is glass and the cover material air. The structure parameters are again similar to the 'Air' and 'Bur' structures. The light line index for the SOI structure is $n_{\parallel} = 1.44$. The band diagram of the 'SOI' structure is

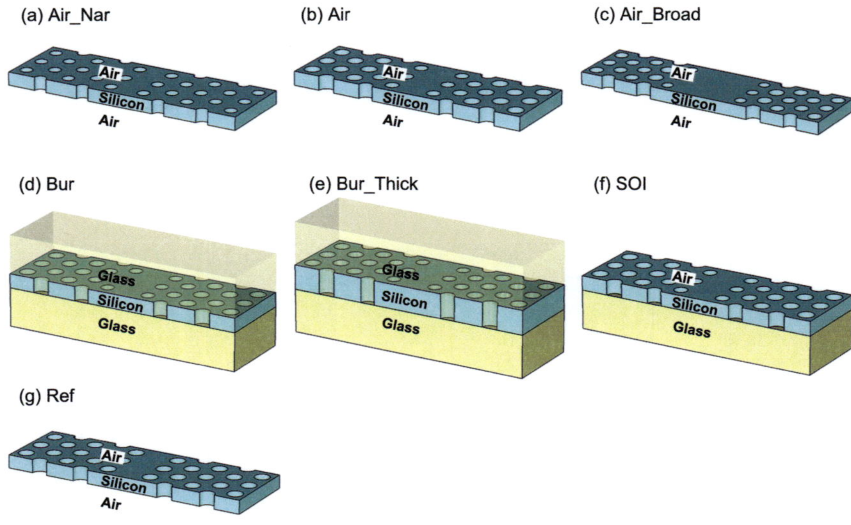

Fig. 2.16. Structure schematics for the broadband slow light designs of Table 2.1.

shown in Fig. 2.17(d), where the TE and TM modes can not be separated any more because of the vertical structure asymmetry. It can be seen that an intrinsic interaction between the previous TE and TM modes occurs, as explained in Section 1.4, which leads to the formation of mini-stop bands.

The group velocity behavior for the different structures is shown in Fig. 2.18. It can indeed be observed that all designs except for the 'Ref' structure have a broadband slow light behavior and a group velocity of 3.7 % of the vacuum speed of light at 193.4 THz. The 'Air', 'Bur', 'Bur_Thick' and 'SOI' structures show an almost identical group velocity behavior. The group velocities of the membrane structures 'Air_Nar', 'Air' and 'Air_Broad' having different core widths start to differ for frequencies above 193.4 THz, where the structures with higher core widths have higher group velocities.

2.2.4   Loss Simulation Results

The losses of the various designs presented in the previous subsection will be determined and discussed in this subsection. In current fabrication processes for silicon PCs, the position of the holes can be controlled to high accuracy, whereas the hole radius and the hole shape might be imperfect. For this reason, we concentrate on the numerical investigation of radial disorder, and we choose a normal distribution of the hole radii about their nominal values. The level of disorder used for all simulations is $\sigma = 5$ nm. The standard deviation σ of the normal distri-

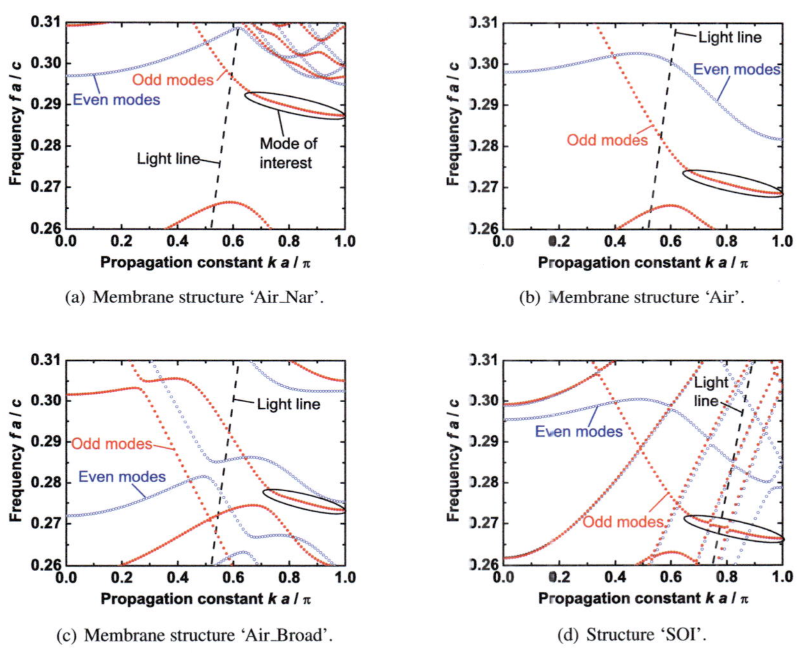

(a) Membrane structure 'Air_Nar'.

(b) Membrane structure 'Air'.

(c) Membrane structure 'Air_Broad'.

(d) Structure 'SOI'.

Fig. 2.17. Band diagrams of broadband slow light structures 'Air', 'Air_Nar', 'Air_Broad' and 'SOI' of Table 2.1. Odd modes are shown as red filled circles, even modes as blue open circles, and the mode of interest is marked by an oval. (c) The structure 'Air_Broad' is multi-moded, and in addition to the mode of interest, four other mode branches exist in the operating frequency range. (d) Due to the vertical asymmetry of the 'SOI' structure, the previously separable TE and TM modes interact leading to mini-stop bands inside the operating frequency range. In parts (a)-(c), only the TE modes are displayed.

bution is the square root of the variance, and σ corresponds to the root mean square (RMS) value of the deviation. This disorder level has been chosen such that it is on the one hand high enough to be resolved by the discretization of the fully numerical FIT method, and that it is on the other hand still suitable for the perturbative approach of the GME method. The best current fabrication processes have a disorder RMS of around 1 nm [103]. The results obtained for a certain disorder level can be translated to other disorder levels by considering the quadratic dependence $\alpha \propto \sigma^2$, which we have found to be valid also for a slow-light test structure. For each simulated structure, various disorder realizations are considered and averaged in order to be able to draw meaningful conclusions. From the obtained transmission and reflection results, the length-independent loss parameters α_b, α_t are determined using the loss model of Subsection 2.2.1.

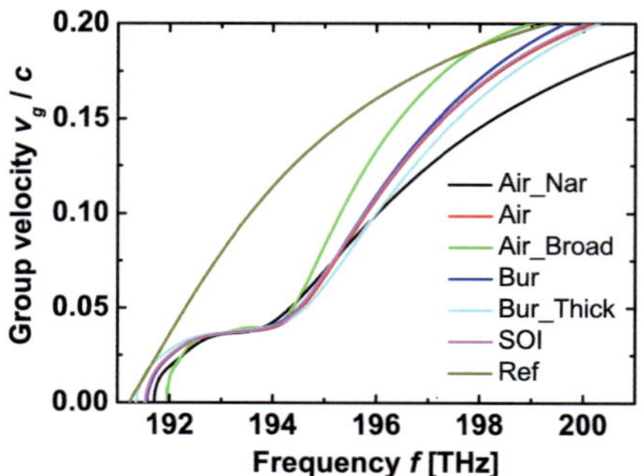

Fig. 2.18. Group velocity for broadband slow light designs of Table 2.1. All designs except the reference structure exhibit a broadband slow light behavior with a group velocity of 3.7 % of the vacuum speed of light at 193.4 THz.

FIT Method

We evaluate the losses of four slow-light designs using the FIT method. Because of the long simulation times it is not possible to simulate all suggested designs of the previous subsection, therefore we restrict ourselves to the two single-moded WGs 'Air' and 'Bur', the multi-moded WG 'Air_Broad' and the vertically asymmetric structure 'SOI'. Significant mode-coupling effects are expected for the 'Air_Broad' and 'SOI' structures, which can be considered using the FIT method. The length of the disordered section is 15 periods, and to efficiently couple to the slow light mode with a strip-WG, a PC taper with a length of 5 periods and 2 additional transition periods on each side of the taper was introduced at the beginning and at the end of the disordered section. To remove the influence of these coupling sections from the transmission and reflection results, a numerical calibration technique is used. This is the transmission-reflect-line (TRL) calibration; more details on this technique are presented in Subsection 4.3.2, and the mathematical background can be found in Appendix A.4. A number of 25 different disorder realizations were simulated for each of the structures, and the transmission and reflection results were averaged over the realizations. The calibrated and averaged transmission characteristics are shown in Fig. 2.19. The solid lines represent the averaged transmission with disorder, whereas the dashed lines show the transmission for the ideal structures. It is to be noted that the length $15\,a$ of the PC-WG is different for the different structures, as the period a is not the same for all designs, Table 2.1. For the ideal structures, the transmission below the light line should be near 0 dB, which can indeed be observed for the 'Air', 'Air_Broad' and 'Bur' structures. In certain spectral ranges, however, the transmission rises to values above 0 dB, which is caused by numerical errors; this can be especially seen for the 'Air_Broad' and 'Bur' structures around 194 THz. The light-line intersection point for the 'Bur' structure can be localized near 194.3 THz. For lower frequencies, the mode is below the light line, while for higher frequencies the mode becomes leaky and the ideal transmission drops. For the 'SOI' structure, however, the ideal transmission below the light line is much lower than for all other structures,

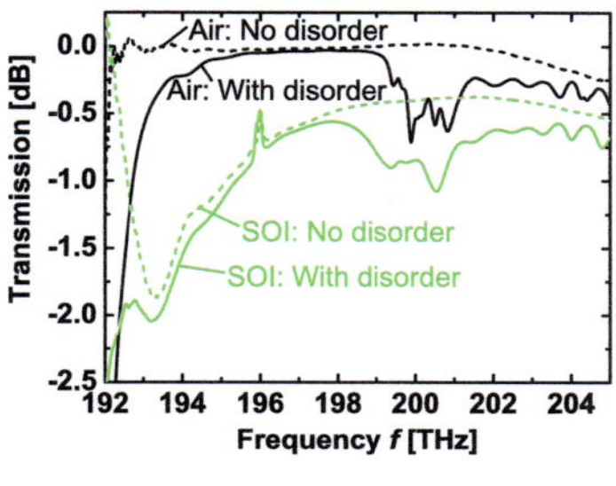

(a) Transmission for 'Air' and 'SOI'.

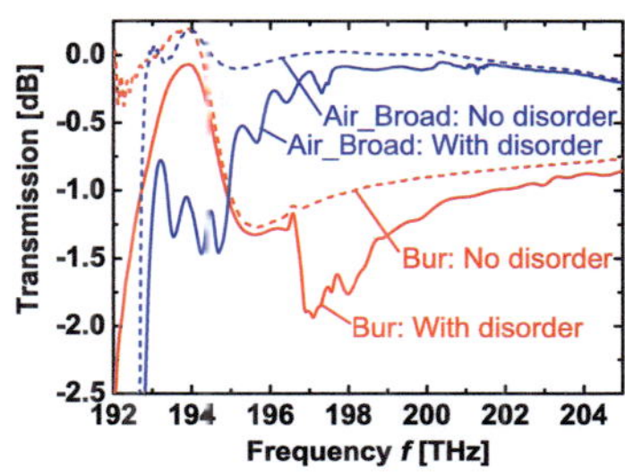

(b) Transmission for 'Air_Broad' and 'Bur'.

Fig. 2.19. Transmission with and without disorder from FIT-simulations for structures 'Air', 'Bur', 'Air_Broad' and 'SOI' of Table 2.1. (a) For 'Air', the ideal transmission is near 0 dB. For 'SOI', the transmission without disorder is already low, which indicates intrinsic TE-TM mode coupling. (b) For 'Bur', the transmission drop caused by the light-line intersection near 194.3 THz is observed. The ideal transmission for both 'Bur' and 'Air_Broad' is larger than 0 dB indicating numerical errors.

and the disorder reduces the transmission only marginally; also the light-line intersection cannot be observed. This indicates that the 'SOI' structure exhibits very large intrinsic losses, and we propose that these are caused by TE-TM coupling which exists even without disorder.

The group velocity curves for the structures with disorder are shown in Fig. 2.20. The influence of the disorder on the group velocity in the broadband slow-light region is only slight which implies that the broadband slow light characteristics are not significantly impeded. Only above 196 THz, oscillations arise at certain frequencies in the spectrum, which can be attributed to mode coupling. It is to be noted that the group velocity characteristics from the FIT method deviates somewhat from the group velocity obtained from GME, Fig. 2.18, which is most likely

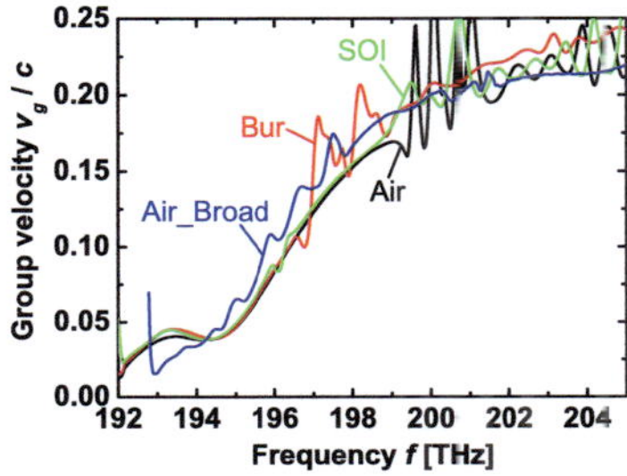

Fig. 2.20. Group velocity for disordered structures 'Air', 'Bur', 'Air_Broad' and 'SOI' of Table 2.1 from FIT simulations. The disorder affects the group velocity only slightly, and the broadband slow light behavior persists. Only above 196 THz, oscillations due to mode coupling occur. For the 'Air_Broad' structure, the group velocity deviates from what is expected.

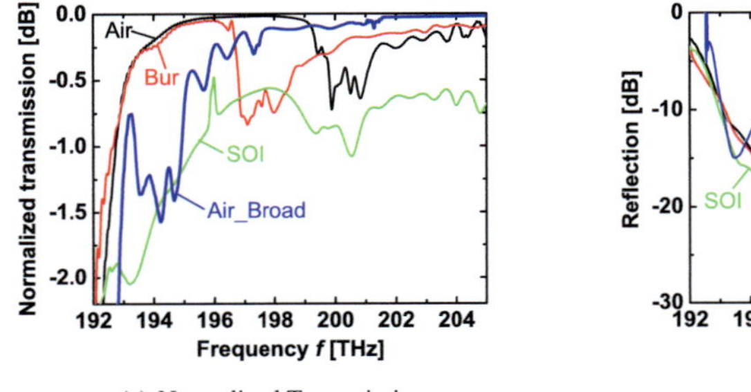

(a) Normalized Transmission.

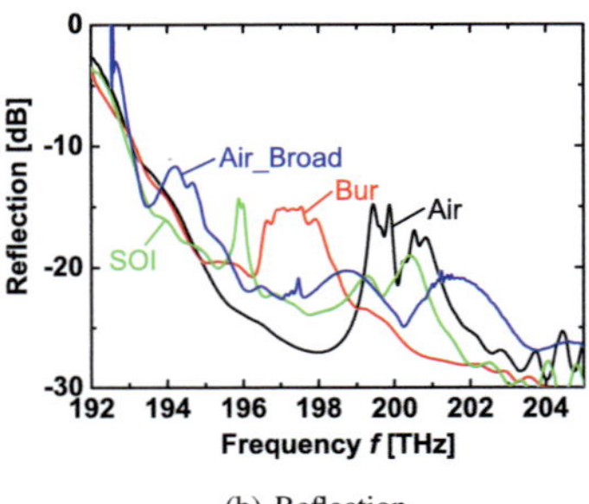

(b) Reflection.

Fig. 2.21. Normalized transmission and reflection for disordered structures from FIT-simulations. The transmission for the 'Air', 'Air_Broad' and 'Bur' structures is normalized to the ideal transmission, while the transmission for 'SOI' is not normalized as intrinsic losses are dominant. Transmission drops and reflection increases due to mode coupling to higher-order modes. (a) Mode coupling above 199.5 THz for 'Air', and above 197 THz for 'Bur'. (b) Multi-moded 'Air_Broad' structure shows mode coupling in whole slow-light region, while 'SOI' has dominant intrinsic coupling.

caused by an insufficient number of modes or plane waves considered in the GME design process, or more generally by the limited accuracy of the methods.

To remove the numerical inaccuracies leading to transmission values larger than 0 dB and the residual in-plane losses, the transmission with disorder is normalized to the ideal transmission for the 'Air', 'Air_Broad' and 'Bur' structures, Fig. 2.21(a). This normalization, however, also subtracts the intrinsic losses that occur for operation above the light line. For the 'SOI' structure the transmission is not normalized, as the intrinsic losses are dominant over the extrinsic losses and normalization would obscure the result. The reflection for all structures is plotted in Fig. 2.21(b). It can be observed that in certain spectral regions, the transmission drops and the reflection increases, which is due to mode coupling to the higher-order TE mode. For the 'Air' structure, the coupling occurs for frequencies above 199.5 THz, for 'Bur' above 197 THz, and for 'Air_Broad' in the complete slow-light region owing to its multi-moded behavior. For the 'SOI' structure, the coupling to higher-order modes is unimportant compared to intrinsic coupling. It can be observed in Fig. 2.20 that mode-coupling leads to oscillations in the group velocity characteristics. Finally, we infer the total losses α_t, the backscattering losses α_b and the out-of-plane losses losses α_o according to the loss model presented in Subsection 2.2.1 from the normalized transmission and reflection data as shown in Fig. 2.21. This is achieved by a nonlinear parameter fit. The loss coefficients α are a better measure for losses as they are in contrast to transmission and reflection independent of length, and are directly comparable. The obtained loss characteristics α_t and α_b are shown in Fig. 2.22. We find that for the four evaluated structures, the results from the model Eqs. (2.3)-(2.5) and from the simplified model Eqs. (2.8)-(2.9) agree very well, which means that the simplified model is sufficient. The transmission is then found by the simple and well-known relation $P_t(L) = \mathrm{e}^{-\alpha_t L}$, and the relevant loss parameter is the total loss α_t.

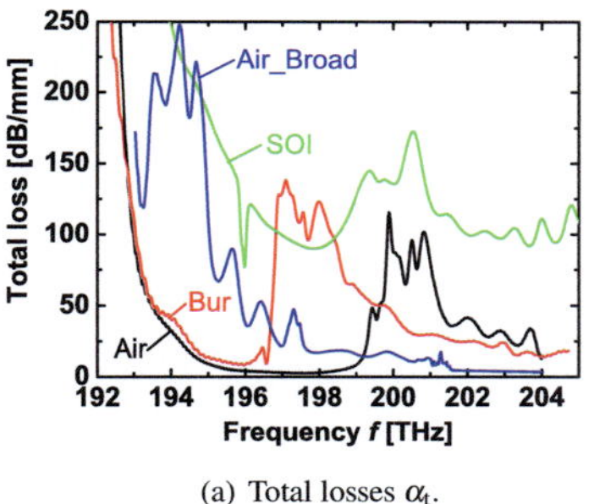

(a) Total losses α_t.

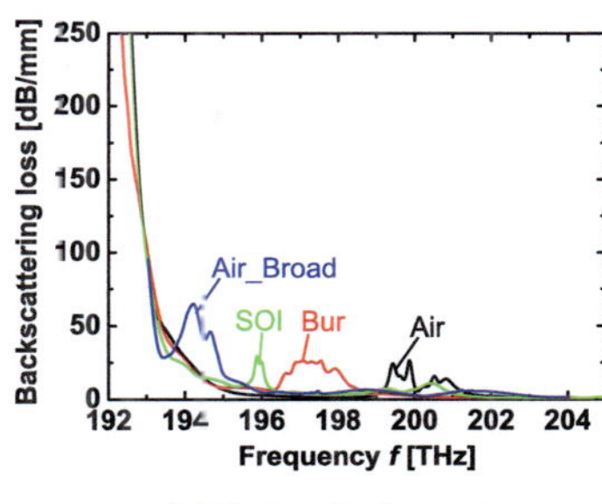

(b) Backscatter losses α_b.

Fig. 2.22. Loss parameters for broadband slow light WGs from FIT. (a) The 'Air' and 'Bur' structures show lowest losses, with 'Bur' slightly higher losses than 'Air'. 'Air_Broad' and 'SOI' have significantly higher losses than 'Air' and 'Bur' because of strong mode coupling. (b) The low-loss structures 'Air' and 'Bur' show dominant backscattering losses at low group velocity.

The FIT simulations predict that the 'Air' and the 'Bur' structures exhibit the lowest total losses, both structures have almost the same losses below the light line, Fig. 2.22(a). The 'Air_Broad' and 'SOI' structures, however, show significantly higher losses, which is due to mode coupling. The backscattering losses are in the same order for all four structures, see Fig. 2.22(b), but the spectral regions of strong mode-coupling can also be observed to increase reflection. For the 'Air' and 'Bur' structures exhibiting low coupling losses between different modes in the slow-light regime, the backscattering losses become the significant contribution of the total losses, Fig. 2.22(b).

GME Method

By using the GME method, the loss parameters of PC-WGs can be determined in much shorter simulation times than by using the FIT method, however in the GME simulations, mode coupling cannot be considered and the validity of the results has to be carefully examined. Thus, we simulate the losses of all single-moded and vertically symmetric broadband slow-light designs of Subsection 2.2.3, which are the structures 'Air_Nar', 'Air', 'Bur', 'Bur_Thick', and of the reference structure 'Ref'. For each structure, the length of the disordered PC-WG was varied between 10 periods and 20 periods in steps of one period, and for each length 10 different disorder realizations were simulated. Thus, the averaging is performed over a total number of 110 realizations. The results for the total losses α_t and the backscattering losses α_b are displayed in Fig. 2.23, where the total losses are the sum of out-of-plane and backscattering losses, $\alpha_t = \alpha_o + \alpha_b$. From the curves of the total loss, Fig. 2.23(a), the intersection point of the modes with the light line can be observed, which is the frequency at which the losses increases again. For the 'Bur' structure, the light-line intersection occurs at around 194 THz, whereas the intersection moves to a higher frequency for the buried structure with larger slab height, 'Bur_Thick'; this is an advantage of a larger slab height. For frequencies above the light line, unphysical jumps in the loss characteristics occur at certain frequencies, and in order to

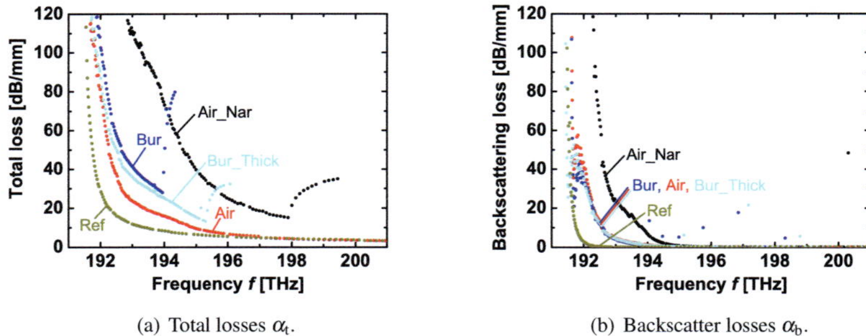

(a) Total losses α_t.   (b) Backscatter losses α_b.

Fig. 2.23. Losses for broadband slow light WGs 'Air_Nar', 'Air', 'Bur', 'Bur_Thick', and 'Ref' from GME method. (a) The slow-light design 'Air' has lowest loss, followed by 'Bur_Thick', 'Bur' and 'Air_Nar'. Thus, the membrane structures have lower losses than the buried structures, a higher WG core width and a higher slab decrease the losses. The 'Bur' and 'Bur_Thick' structures of higher cladding index exhibit decreased low-loss range below the light-line, and 'Bur_Thick' with higher slab has increased range (b) Backscattering loss for all structures is much smaller than total loss.

also obtain reliable results above the light line the number of averages would need to be further increased. It can be observed from the graphs that the backreflection loss for all designs is predicted to be much lower than the total loss in the frequency range of interest above approximately 192 THz, which suggests that the total loss is mainly determined by out-of-plane loss.

Clearly, the reference structure 'Ref' shows lowest losses, as it exhibits the highest group velocity. Comparing the slow-light membrane structures of different WG core width 'Air_Nar' and 'Air' reveals that for larger core width, the losses are significantly smaller. The designs 'Air' and 'Bur' mainly differ in their cladding material but have otherwise very similar structure parameters, and the 'Air' structure having lower cladding index than the 'Bur' structure is predicted to have smaller losses. Comparing the two buried structures 'Bur' and 'Bur_thick' of different silicon slab thickness, the design with larger slab height 'Bur_thick' has slightly lower losses.

Comparison

For the two single-moded and vertically symmetric structures 'Air' and 'Bur', the losses are obtained from both the GME and FIT methods and can be compared, Fig. 2.24. A slight frequency shift is introduced in the GME curves in order to match the group velocity characteristics of Fig. 2.18 to that of Fig. 2.20. The solid lines represent the losses obtained from FIT, and the dashed lines the results from GME. The losses from the FIT simulations are normalized to the ideal structure and thus represent the disorder-induced excess losses, as described above. The results from GME are truncated for frequencies above the light line intersection point as unphysical jumps appear. It can be seen that the total losses obtained from both methods show a qualitatively similar behavior, but the FIT method predicts higher losses for propagation at low

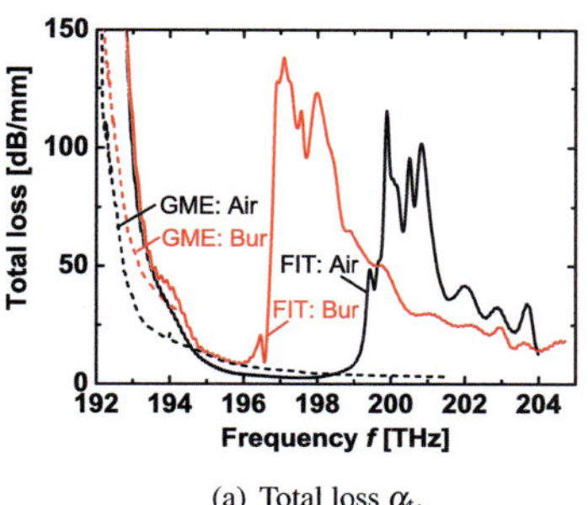
(a) Total loss α_t.

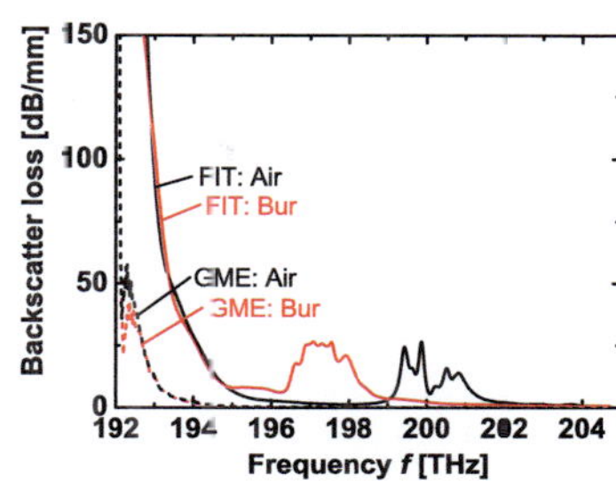
(b) Backscattering loss α_b.

Fig. 2.24. Comparison of losses obtained from GME and FIT for 'Air' and 'Bur' structures. The FIT method predicts higher total loss at low group velocity and much higher backscattering loss than the GME method. With both methods, 'Air' has lower losses than 'Bur'.

group velocity. At the same time, the reflection losses from the GME method are predicted to be much lower than from the FIT method.

As discussed above, we assume the FIT method is more reliable than the GME method in predicting losses, as the FIT method can model all loss effects and does not use any approximations except for a limited structure discretization and simulation time. Furthermore, the loss results from FIT are verified by results from FDTD for test structures, see Section 4.3. Although the results from the GME method do not agree perfectly, we suggest that the method is well suited to compare the losses of PC-WGs that are single-moded and exhibit no significant mode-coupling, and because of its low simulation times it is ideal for structure optimization.

2.2.5 Dependance of Losses on Group Velocity

The disorder-induced losses of a PC mode are strongly dependent on its group velocity. The losses increase for decreasing group velocity, which can be observed in Fig. 2.22, where the group velocity decreases for decreasing frequencies. The increased losses at low group velocity can be intuitively understood, as for a slower optical wave its interaction time with the scatterers is longer, or the possibility of a slow photon bouncing many times 'back and forth' is higher to be scattered.

In three different works, the dependence of extrinsic losses with group velocity has been previously studied; some of the findings have already been mentioned in the introduction of this section. In [22], it has been identified that the propagation loss α is approximately inversely proportional to group velocity v_g, $\alpha \propto \frac{1}{\lambda v_g}$. In a physical model using Green's functions [58] it has been predicted and inferred from measurements [59] that the out-of-plane scattering loss scales inversely proportional with group velocity, $\alpha_o \propto \frac{1}{v_g}$, and the backscattering loss scales inversely proportional with the square of the group velocity, $\alpha_b \propto \frac{1}{v_g^2}$; for low group velocity, the backscattering loss dominates. In [103] it was suggested that out-of-plane scattering is dominant for group velocities higher than 5 % of the vacuum speed of light c and scales sub-linearly with group velocity, $\alpha_o \propto \frac{1}{\sqrt{v_g}}$, whereas backscattering with inverse proportionality to the square of the group velocity only becomes dominant for group velocities lower than $1\,\%\,c$. It is suggested that near the band-edge, reflections caused by local variations of the lattice become the dominant source of loss.

The results of these works are partly inconsistent. Furthermore, the studied PC-WGs in these works exhibit a monotonously increasing group velocity for increasing frequency without any characteristic shape, and the group velocity characteristics have not been measured but are numerically determined and matched to the experiments. However, it is not fully known if the group velocity match is correct; the fabricated structures may deviate from the simulations, and a frequency offset may have occurred. Also, the fit of a monotonous group index curve to a monotonous loss curve, both without any characteristic features, might be possible even if the interdependence is rather weak.

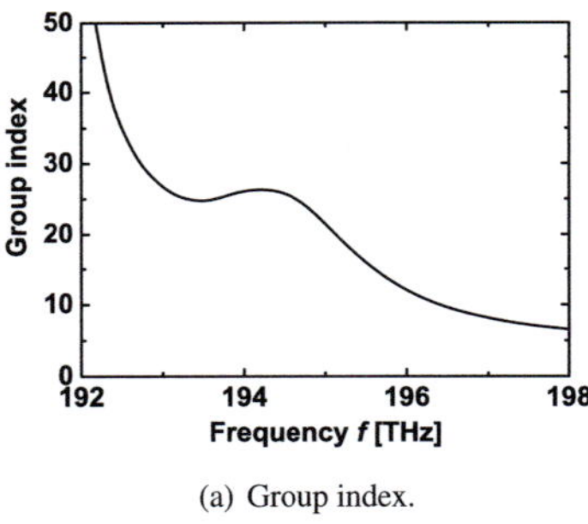

(a) Group index.

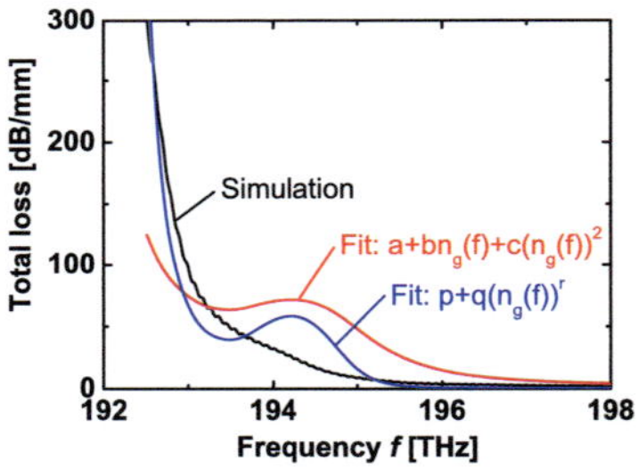

(b) Total loss α_t and best fits with group index n_g.

Fig. 2.25. Dependance of total losses on group index for 'Air' structure. (a) The group index has a constant value in the slow-light region. (b) Total loss and best quadratic and power fits as function of group index. The group index characteristics cannot be observed in the loss spectrum, and neither of the fits is correct.

For our broadband slow-light structure under study, the group velocity is constant in a certain spectral range, which represents a characteristic feature that should also appear in the loss spectrum if a strong dependance is present. We exemplarily examine the FIT results of the 'Air' structure; its group index n_g and total losses α_t are shown in Fig. 2.25. Two different fits for the total losses as function of group index are shown in the graph, which are a quadratic fit of the form $\alpha_t(n_g(f)) = a + bn_g(f) + c\left[n_g(f)\right]^2$ and a power fit of the form $\alpha_t(n_g(f)) = p + q\left[n_g(f)\right]^r$, where the coefficients a,b,c,p,q,r are determined from a least square fit. It can be observed that neither of the two fits is satisfactory, and the total loss curve does not show the characteristic feature present in the group velocity. The exponent of the power fit has a value $r = 6.7$ that is far from the value of 2 or 0.5, as previously predicted. Similarly for the other studied slow light structures, we find that its loss characteristics do not reflect the broadband group index or group velocity behavior.

In contrast to previous reports, we cannot verify a quadratic, linear or square root dependance of the disorder-induces losses on the group index.

2.2.6 Conclusion

Various broadband slow-light PC-WGs with a group velocity of 3.7% of the vacuum speed of light are designed, and the disorder-induced losses are numerically evaluated. A loss model is introduced, and two different numerical tools are utilized taking their limitations into account. We find that an air membrane structure with a WG core width as large as possible to be still single-moded has lowest disorder-induced losses. For a buried structure, where glass is added as cladding material, the losses are a slightly larger than for the air membrane structure, and the operation range below the light line is greatly reduced. A larger thickness of the silicon slab can increase the operation range below the light line and can slightly reduce losses. However, multi-moded or vertically asymmetric WGs exhibit strong mode-coupling effects that significantly increase losses; such structures are less suitable to be used as low-loss broadband slow-light WGs. Furthermore, we show that the group velocity characteristics of the slow-light WG can not be identified in the loss spectrum, which is contradictory to previous findings that suggest a direct dependance of the losses on the group velocity.

Chapter 3

Devices

Numerous powerful functional devices in chip-scale have been proposed on the basis of photonic crystals. These devices might be passive, like high-Q cavities [106], [107], splitters [108], filters [109], or multiplexers and demultiplexers [110], or they might also be nonlinear or active devices like lasers [111], [112] or switches [113].

In this chapter, we primarily focus on novel passive devices that benefit from the unique slow-light or highly-dispersive properties of PC-WGs. The application of the devices is especially for optical signal processing in modern telecommunication networks at high bit rates of 40 Gbit/s or above. In the previous chapter, we have developed detailed knowledge about broad-band slow-light PC-WGs, which will be used in this chapter for the design of functional devices. In Section 3.1, we present a tunable dispersion compensator, in Section 3.2 a tunable delay line, and in Section 3.3 a fast electro-optic modulator. All three devices employ one of the tuning mechanism mentioned in Section 1.7 to be able to control the device behavior. In addition to the device designs presented here, measurements of actually fabricated basic PC-WGs will be presented in Sections 4.2 and 4.1.

3.1 Tunable Dispersion Compensator

In high-capacity optical communication systems with high bit rates, the compensation of chromatic dispersion accumulated in the fiber link is essential. The higher the bit rate and thus the signal bandwidth, the more severe is the signal degradation by dispersion, as the dispersion-induced temporal pulse spread is proportional to the bandwidth $|\Delta\lambda|$, $|\Delta t_{\mathrm{g}}| = CL\,|\Delta\lambda|$, see Eq. (1.25). In most cases, dispersion is compensated using dispersion compensating fibers (DCFs) having negative chromatic dispersion, where each link of standard SMF is followed by a coil of DCF of appropriate fiber length. The DCFs are, however, expensive, introduce considerable additional losses and nonlinearities, and are fixed in length, which does not allow changing the dispersion value after installation. To this end, numerous other schemes are proposed to deal with dispersion. The first direction is to completely obviate optical dispersion compensation, e. g. by developing and improving modulation formats that are more tolerant to dispersion, like optical orthogonal frequency division multiplexing [114], by demultiplexing fast data streams into several channels with lower rates, or by using electronic dispersion compensation [115], [116], [117]. The other direction is to develop compact and adjustable optical dispersion compensators, which can be realized by chirped fiber Bragg gratings [118], virtually imaged phased arrays [119], all-pass multicavity etalons [120] or by all-pass filters based on a micro-electro-mechanical system (MEMS) [121]. Furthermore, optical dispersion compensators can also be realized in integrated planar waveguide circuit, e. g. with coupled strip waveguides [122], all-pass ring resonator filters [123], all-pass Mach-Zehnder interferometers [124], waveguide grating routers [125] or with phase-conjugating silicon waveguides [126].

Photonic crystals show in particular very strong dispersive properties: PC fibers exhibit a much larger negative chromatic than standard DCFs [127], and integrated PC-WGs show regions of extremely high chromatic dispersion [14]. Based on such PC-WGs, compact devices for dispersion compensation have been suggested, [39], [128], [129].

In this section, we propose a tunable dispersion compensator based on two concatenated PC-WGs that exhibits a flat dispersion in a wide 125 GHz bandwidth. The dispersion can be continuously tuned in a range between $-19\,\mathrm{ps}/(\mathrm{mm\,nm})$ and $7\,\mathrm{ps}/(\mathrm{mm\,nm})$, and the total length of the device determines the dispersion-length product and thus the dispersion that can be compensated. The device comprises two concatenated PC-WGs sections, where one of the sections has a linearly increasing chromatic dispersion characteristics as a function of frequency, while the other one has a linearly decreasing dispersion. The PC-WGs sections can be designed such that a flat total dispersion is obtained, and tuning one of the sections changes the dispersion value.

The section is organized as follows: In Subsection 3.1.1, the operation principle of the device is explained. In Subsection 3.1.2, details about the PC-WG design are shown, and in Subsection 3.1.3, simulation results of the complete device are presented indicating its feasibility.

3.1.1 Principle of Operation

The total chromatic dispersion C_{tot} of two concatenated WG sections with chromatic dispersion characteristics $C_{1,2}$ is calculated according to

$$C_{\mathrm{tot}} = \frac{C_1 l_1 + C_2 l_2}{l_1 + l_2}, \tag{3.1}$$

where the two sections have lengths $l_{1,2}$. Multiplying the dispersion C_{tot} with the total length of the structure, $L_{\mathrm{tot}} = l_1 + l_2$, the dispersion-length product $C_{\mathrm{tot}}L_{\mathrm{tot}}$ is obtained, and obviously increasing the total length L_{tot} of the device increases the dispersion that can be compensated, while the ratio of l_1 and l_2 needs to stay fixed in order to maintain the dispersion characteristics.

The tunable chromatic dispersion compensator is based on a PC-WG that is specifically designed to have a dispersion characteristics as shown in Fig. 3.1(a), which has already been mentioned in Subsection 2.1.3. The chromatic dispersion characteristic exhibits a spectral region with a linearly decreasing value, then the minimum dispersion is reached having a negative value, and a region with a linearly increasing dispersion follows. The linear dispersion regions are marked by ovals in Fig. 3.1(a). A second section of PC-WG is introduced and is appropriately shifted in frequency by slightly changing the structure parameters, so that the regions of linearly increasing and decreasing dispersion of sections 1 and 2 overlap, see Fig. 3.1(b). The ratio of the lengths of the two sections can be adjusted such that the dispersion-length products

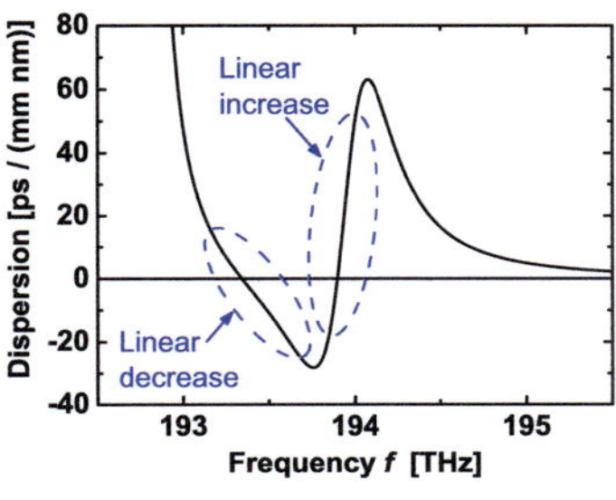
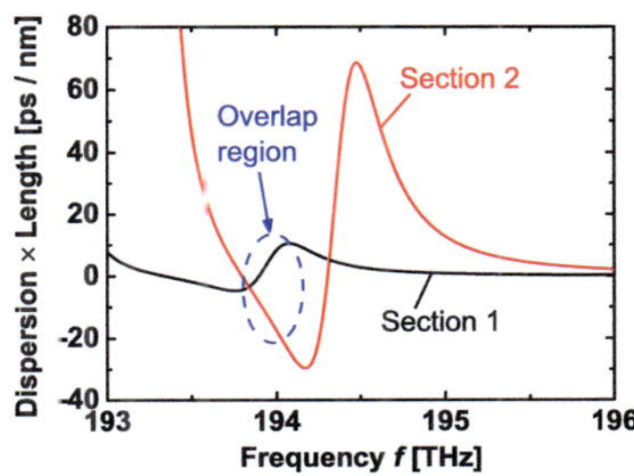

(a) Dispersion characteristics of PC-WG.   (b) Dispersion-length product of PC-WG sections.

Fig. 3.1. Chromatic dispersion characteristics of PC-WG used for dispersion compensator. (a) The specifically designed PC-WG exhibits two spectral regions with linearly varying dispersion and negative minimum dispersion. (b) By concatenating two PC-WG sections of appropriate length and by frequency-shifting one of the sections, the two linear regions overlap and a flat total dispersion results.

in the overlap region have equal slopes but opposite sign, which means that the concatenation of the two sections results in a total dispersion characteristic with a nearly constant value in a considerable bandwidth. The schematic of the device structure is shown in Fig. 3.2, where coupling tapers have been introduced at the beginning and end of the PC-WGs to increase the transmission, as explained in Subsection 2.1.4.

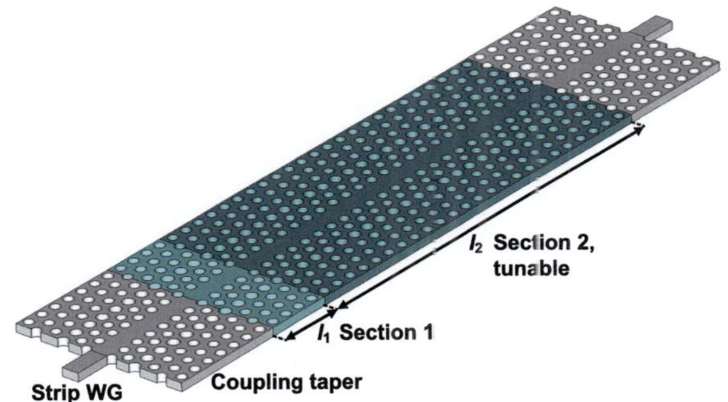

Fig. 3.2. Structure schematic of tunable dispersion compensator. Two sections of lengths l_1 and l_2 are concatenated, where section 1 is operated in the range of linearly increasing and section 2 of linearly decreasing dispersion. If section 2 is tuned, its characteristics is shifted in frequency, and the resulting total dispersion can be varied. To increase the coupling to strip-WGs, PC tapers are introduced at both ends of the structure.

The resulting dispersion value of the device can now be made tunable by shifting the dispersion characteristics of section 2 in frequency, which can be achieved with one of the tuning mechanisms introduced in Section 1.7, e. g. with free carrier injection or thermal tuning. The spectral range of linearly increasing dispersion of section 1 represents the maximum bandwidth of the flat total dispersion, and the maximum tuning range of section 2 is limited by the fact that its linearly decreasing dispersion always needs to overlap with the linearly increasing dispersion of section 1 in order to result in a flat total dispersion. Optical bandwidths of more than 125 GHz can be reached, which makes the device appropriate for transmission at 40 Gbit/s or faster.

3.1.2 PC-WG Design

Numerous parameter sets are possible to obtain a PC-WG with a dispersion characteristic as in Fig. 3.1(a); here we present three designs that exhibit different values of the minimum dispersion and also different group velocity. The group velocity and chromatic dispersion of the designs obtained by the GME method are shown in Fig. 3.3. The designs 1 and 2 are membrane structures, while design 3 is a structure with glass cladding; all three structures have a slab thickness of $h = 220$ nm. The dispersion minimum of design 1 is -28.2 ps$/$(mm nm), of design 2 it is -12.8 ps$/$(mm nm), and of design 3 it is -3.4 ps$/$(mm nm). We find that a higher negative dispersion value can be reached by simultaneously decreasing W_2 and increasing W_3, while the linearity of the decreasing dispersion region can be enhanced by slightly increasing r_1. These three parameters have been changed from design 2 to design 1. Structure 3 has also been fabricated and measured, see Section 4.1. The glass cladding, however, reduces the available operating range below the light line, and the light line intersection occurs already at 193.34 THz, which means that only the linearly decreasing dispersion region is below the light line. The parameters of design 1 are $W_1 = 0.9\sqrt{3}a$, $W_2 = 0.8\frac{1}{2}\sqrt{3}a$, $W_3 = 0.9\frac{1}{2}\sqrt{3}a$, $r_1 = r_2 = r = 0.3a$, $r_3 = 0.35a$, and $a = 460$ nm. The parameters of design 2 are $W_1 = 0.9\sqrt{3}a$, $W_2 = 0.83\frac{1}{2}\sqrt{3}a$, $W_3 = 0.87\frac{1}{2}\sqrt{3}a$, $r_1 = 0.28a$, $r_2 = r = 0.3a$, $r_3 = 0.35a$, and $a = 452$ nm, and the parameters of design 3 are $W_1 = 0.95\sqrt{3}a$, $W_2 = \frac{1}{2}\sqrt{3}a$, $W_3 = 0.9\frac{1}{2}\sqrt{3}a$, $r_1 = 0.25a$,

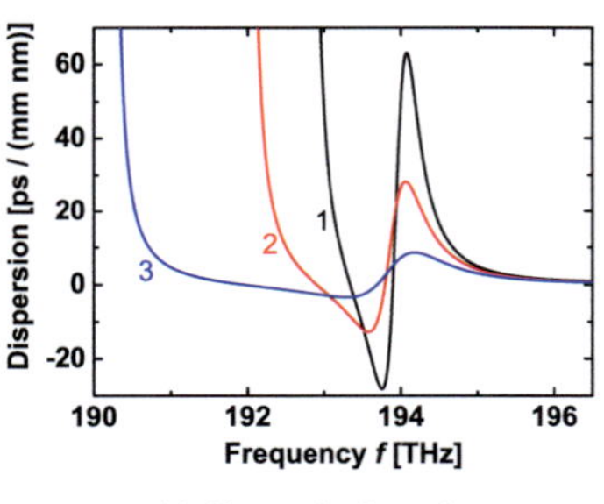

(a) Chromatic dispersion.

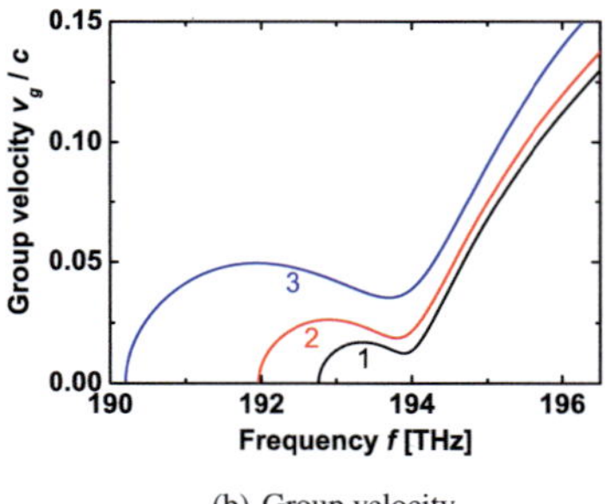

(b) Group velocity.

Fig. 3.3. PC-WGs for tunable dispersion compensator. (a) Design 1 has a minimum chromatic dispersion of -28.2 ps$/$(mm nm), design 2 of -12.8 ps$/$(mm nm), and design 3 of -3.4 ps$/$(mm nm). For lower dispersion, the bandwidth of the linear regions is larger. (b) The group velocity decreases with higher negative dispersion. Structures 1 and 2 are silicon membrane structures, and structure 3 has a glass cladding. The structure parameters are mentioned in the text.

$r_2 = 0.32\,a$, $r_3 = 0.34\,a$, $r = 0.3\,a$, and $a = 410\,\text{nm}$. In general, if the dispersion minimum of the PC-WG is more negative, the dispersive device can be made smaller to reach a given dispersion-length product. However, at the same time the group velocity is decreased, which increases the disorder-induced losses, as discussed in Section 2.2, and the bandwidth of both the linearly increasing and linearly decreasing dispersion regions is decreased. Thus, a compromise needs to be found for a device with a specified bandwidth and tuning range for the dispersion-length product.

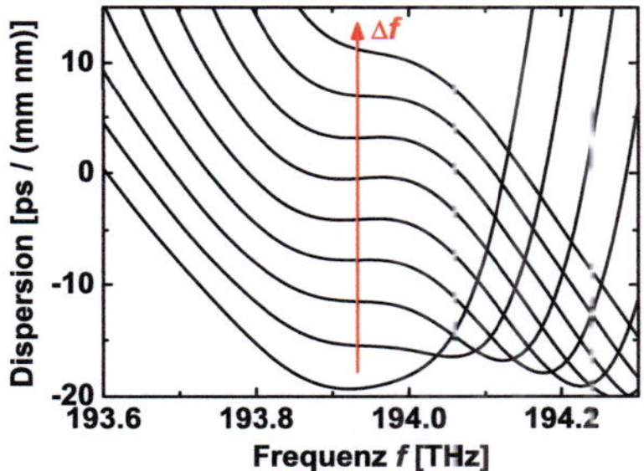

Fig. 3.4. Total dispersion of tunable dispersion compensator with varying tuning Δf of section 2. In a bandwidth of 125 GHz, a chromatic dispersion between $-19\,\text{ps}\,/\,(\text{mm nm})$ and $7\,\text{ps}\,/\,(\text{mm nm})$ can be obtained. The required frequency tuning range of section 2 is $\Delta f = 0.5\,\text{THz}$.

We use design 1 as the basic structure of the tunable dispersion compensator, and the achievable total dispersion characteristics is shown in Fig. 3.4. The different curves correspond to different frequency shifts of section 2 of the PC-WG. The chromatic dispersion can be tuned in a range between $-19\,\text{ps}\,/\,(\text{mm nm})$ and $7\,\text{ps}\,/\,(\text{mm nm})$, and the bandwidth of flat dispersion is as large as 125 GHz. To achieve the full tuning range, the required frequency shift of section 2 is $\Delta f = 0.5\,\text{THz}$. The ratio of the lengths of the two sections needs to be $l_2/l_1 = 7.5$ to reach a flat dispersion. Changing the ratio of l_1 and l_2 to another value allows modifying the resulting dispersion slope to a non-flat dispersion, which might be quite useful for certain applications.

3.1.3 Simulation of Device Characteristics

In addition to the estimated dispersion of the device, we have simulated the transmission and reflection properties of a complete device for a fixed dispersion with the FIT method; the structure is again based on the PC-WG design 1. The simulated device has the structure as shown in Fig. 3.2, and the results are given in Fig. 3.5. The operating frequency observed from the chromatic dispersion graph, Fig. 3.5(b), is slightly shifted with respect to the GME results shown in Fig. 3.4 to about 193.2 THz. The dispersion characteristics obtained from the FIT device simulations show some deviations from the characteristics predicted by the FIT band simulations, which we attribute to a device length that might not be long enough, to the limited frequency

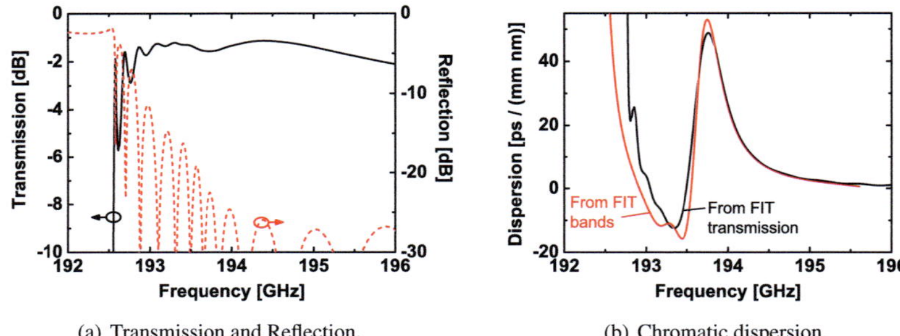

(a) Transmission and Reflection.   (b) Chromatic dispersion.

Fig. 3.5. Transmission, reflection and chromatic dispersion of the dispersion compensator structure as shown in Fig. 3.2. (a) The transmission is better than $-2\,$dB and the reflection below $-10\,$dB in the operating frequency range, which demonstrates an efficient mode excitation (b) A general agreement can be observed between the chromatic dispersion from the FIT device simulation and the prediction by FIT band simulations.

resolution of the simulation, and to the coupling-related oscillations in the transmission. However, the characteristics agree in general. From Fig. 3.5(a) it can observed that the transmission is better than $-2\,$dB and the reflection is below $-10\,$dB in the frequency range of interest, which shows that the mode can be efficiently excited and that the device is feasible.

3.1.4   Conclusion

We show a concept for a tunable optical dispersion compensator, and present a design with a dispersion tuning range from $-19\,$ps$\,/\,$(mm nm) to $7\,$ps$\,/\,$(mm nm) in a 125 GHz bandwidth. The compensator uses two concatenated dispersion-engineered slow-light PC-WGs, where the positive dispersion slope of the first section equalizes the negative dispersion slope of the second section. A complete device simulation predicts coupling losses below $-1\,$dB for each interface from strip-WG to PC-WG, and a reflection below $-10\,$dB and thus shows the feasibility of the device.

3.2 Tunable Optical Delay Line

In next-generation all-optical networks, tunable optical delay lines might play an important role for optical buffering, data synchronization, optical memory and optical signal processing. Furthermore, tunable optical delays are also needed in metrology e. g. for LIDAR (Light Detection and Ranging) or RADAR (Radio Detection and Ranging) applications. The required device bandwidth in fast communication system is in the order of twice the transmitted bit rate, and for 40 Gbit/s transmission an optical bandwidth of 125 GHz is safe. Bandwidths of this order of magnitude cannot be reached using electromagnetically induced transparency (EIT). By using coherent population oscillation (CPO) in a semiconductor, a maximally tunable delay of 2 ps corresponding to 0.07 pulsewidths was shown at 16.7 GHz modulation [76]. By using stimulated Brillouin scattering (SBS), data with a rate of 14 Gbit/s could be delayed up to 0.62 pulsewidths [130]. With stimulated Raman scattering (SRS), an optical signal of a bandwidth as high as 1 THz was delayed by 0.85 pulsewidths [79]. Much higher delays can be achieved using self-phase modulation and dispersion, where a 10 Gbit/s data signal was continuously delayed up to 45 pulsewidths, and maximum data rates of 60 Gbit/s [131] are possible in principle. In integrated optical ring resonators, a 20 Gbit/s optical signal was delayed by a fixed amount of 10 bits [85], and a 2.5 Gbit/s signal was delayed up to 2 bits [132].

In this section, we suggest a tunable optical delay line based on a silicon slow-light photonic crystal. The device is designed to have a bandwidth of constant delay of 125 GHz, and a 40 Gbit/s 33 % RZ signal can be continuously delayed from 0 to 5 pulsewidths on a device length of 1 mm only. The principle of operation is similar to that of the tunable dispersion compensator presented in Section 3.1. Two sections of PC-WG are concatenated, where one section exhibits a positive flat chromatic dispersion and the other section a negative flat dispersion. The resulting group delay is constant in a certain bandwidth, and by tuning one of the sections the total delay can be changed.

This section is organized as follows: In Subsection 3.2.1, the device principle is explained in more detail, followed by PC-WG design considerations in Subsection 3.2.2. An optimization strategy is presented in Subsection 3.2.3, and the tunable delay characteristics are estimated.

3.2.1 Principle of Operation

The group delay t_g in a device or waveguide of length L is related to the group index n_g and to the group velocity v_g by the following equations:

$$t_g = L \frac{1}{v_g} \tag{3.2}$$

$$= L \frac{n_g}{c}, \tag{3.3}$$

where c is the vacuum speed of light. The total group delay $t_{g,\text{tot}}$ of two different concatenated WG section with group delays $t_{g1,2}$ or group indices $n_{g1,2}$ and lengths $l_{1,2}$ is calculated by

$$t_{g,\text{tot}} = t_{g1} + t_{g2} \tag{3.4}$$

$$= l_1 \frac{n_{g1}}{c} + l_2 \frac{n_{g2}}{c}, \tag{3.5}$$

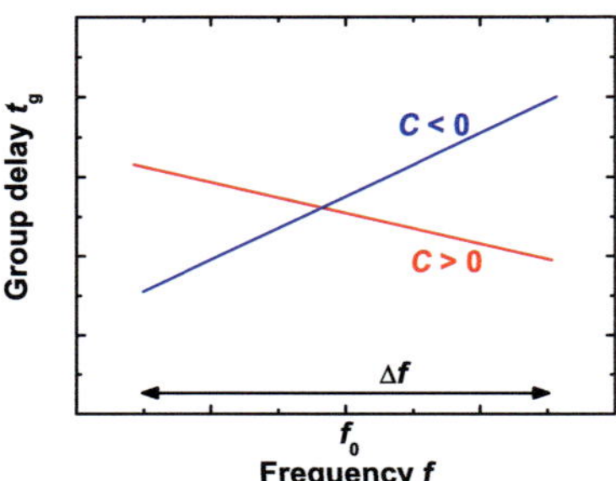

Fig. 3.6. Group delay characteristics of a WG section of constant chromatic dispersion C inside a frequency range Δf. For a positive chromatic dispersion, the group delay decreases linearly with frequency, and for a negative chromatic dispersion, the group delay increases linearly.

where the total length of the device is $L_{\text{tot}} = l_1 + l_2$. The relation between the chromatic dispersion and the group delay can according to Eqs. (1.23)-(1.24) be written as

$$C = -\frac{1}{L}\frac{f_0^2}{c}\frac{\mathrm{d}t_g}{\mathrm{d}f}, \tag{3.6}$$

where the characteristics is regarded in a small fractional bandwidth around a center frequency f_0 (otherwise, f_0 needs to be replaced by f). This means that for a constant chromatic dispersion in a certain bandwidth Δf around f_0, the group delay varies linearly with frequency, see Fig. 3.6, and the slope and sign depend on the chromatic dispersion value and sign. For a

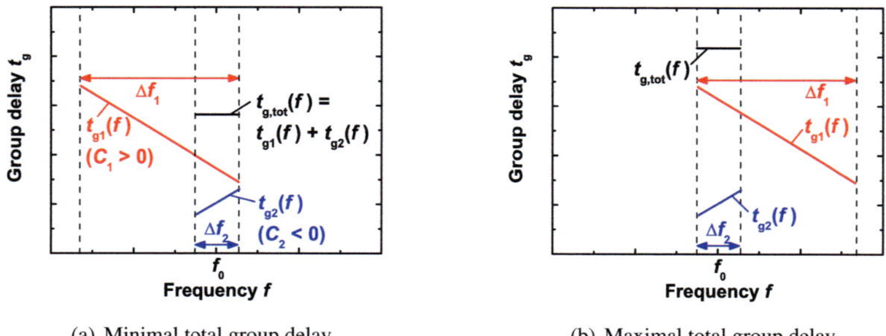

(a) Minimal total group delay.   (b) Maximal total group delay.

Fig. 3.7. Group delay of two concatenated WG sections of constant dispersion. Section 1 has positive chromatic dispersion C_1 in a bandwidth Δf_1, and section 2 has negative chromatic dispersion C_2 in a bandwidth $\Delta f_2 < \Delta f_1$. The total group delay $t_{g,\text{tot}}$ of the concatenated sections is flat. Shifting the characteristics of section 1 in frequency changes the total delay around f_0 from the (a) minimal to the (b) maximal value.

positive chromatic dispersion, the group delay decreases with frequency, while for a negative chromatic dispersion, the group delay increases. If two WG sections of opposite sign of the dispersion are concatenated and the lengths are adjusted properly, a constant total group delay can be reached in a bandwidth which is the minimum of the bandwidths of both sections, see Fig. 3.7(a). Section 1 has positive chromatic dispersion C_1 in a bandwidth Δf_1, and section 2 has negative chromatic dispersion C_2 in a bandwidth Δf_2. For a line-defect PC-WG, the chromatic dispersion of the fundamental mode is positive in the whole frequency range, while careful design can generate rather narrow spectral regions of negative dispersion, see Subsection 2.1.3. Thus, the bandwidth Δf_1 of constant positive chromatic dispersion of section 1 is larger than the bandwidth Δf_2 of constant negative chromatic dispersion of section 2, $\Delta f_1 > \Delta f_2$. By tuning section 1, e. g. by thermal heating or free-carrier injection, its group delay characteristics is shifted in frequency. The total resulting delay $t_{\mathrm{g,tot}}$ of the concatenated sections can then be tuned from a minimal value Fig. 3.7(a) to a maximal value Fig. 3.7(b), while the characteristics is flat in a bandwidth Δf_2 around the frequency f_0. The total structure of the tunable delay line is schematically shown in Fig. 3.8. The coupling from strip-WGs to the PC-WGs having low group velocity is increased by introducing PC tapers at the start and the end of the device; the tapers can be designed according to the principles presented in Subsection 2.1.4.

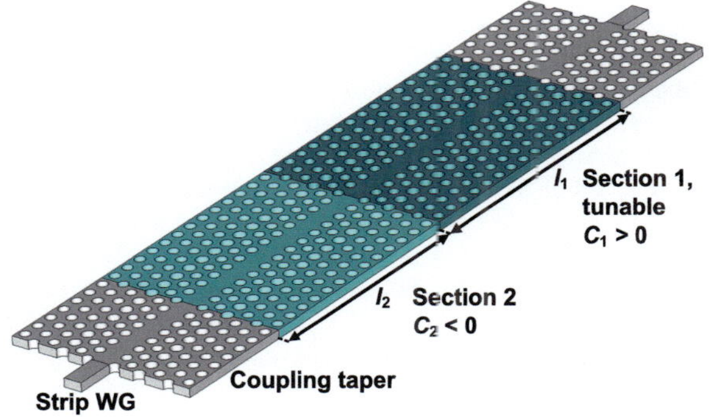

Fig. 3.8. Structure schematic of tunable delay line. Two sections of lengths l_1 and l_2 are concatenated; section 1 has a constant positive chromatic dispersion, and section 2 a constant negative chromatic dispersion. By tuning section 1, the resulting group delay can be varied. To increase the coupling to strip-WGs, PC tapers are introduced at both ends of the structure.

3.2.2 PC-WG Design

Constant Negative Chromatic Dispersion

We begin with the design of section 2 of the tunable delay line, which has a constant negative chromatic dispersion. Such a characteristic is met by the tunable dispersion compensator, which has been designed and presented in Section 3.1. However, we need not tune the dispersion

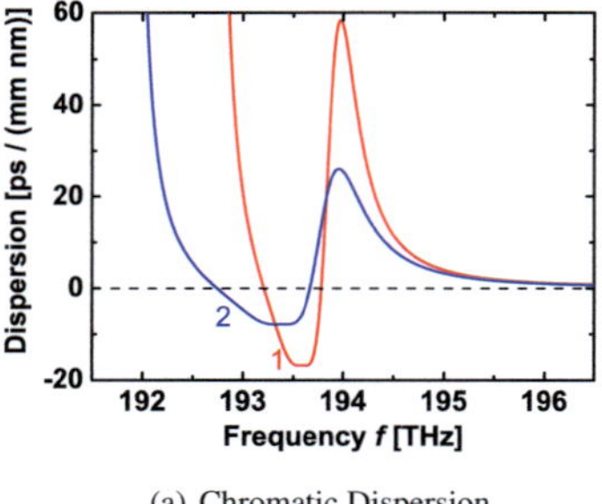

(a) Chromatic Dispersion.

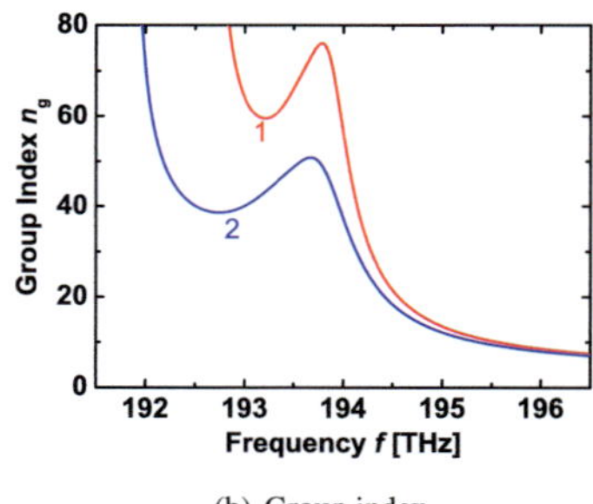

(b) Group index.

Fig. 3.9. PC-WG devices with constant negative chromatic dispersion. Each of the devices consists of two parts of PC-WG that are slightly shifted in frequency to obtain the broadband behavior. (a) Design 1 has a larger negative dispersion $C = -16.8\,\mathrm{ps}/(\mathrm{mm\,nm})$ and a narrower bandwidth $\Delta f = 200\,\mathrm{GHz}$ compared to design 2 with $C = -7.9\,\mathrm{ps}/(\mathrm{mm\,nm})$ and $\Delta f = 320\,\mathrm{GHz}$. (b) The group index n_g increases linearly with frequency in the region of constant dispersion, and the design 1 with larger negative chromatic dispersion has higher group index than design 2. Structure parameters are mentioned in the text.

value any more, but fix the characteristics to the minimum flat dispersion. The constant negative chromatic dispersion has been achieved by concatenating two parts of PC-WG, where one of the parts has a slightly larger WG core width W_1 to shift the dispersion characteristic in frequency without changing the shape. We show two possible designs of different chromatic dispersion value here, which are based on the two membrane structure designs 1 and 2 of the dispersion compensator. The chromatic dispersion and group index of the two designs are displayed in Fig. 3.9. The spectral regions of constant negative dispersion can be clearly observed, and design 1 has a larger negative dispersion of $C = -16.8\,\mathrm{ps}/(\mathrm{mm\,nm})$ and a narrower bandwidth $\Delta f \approx 200\,\mathrm{GHz}$ than design 2 with $C = -7.9\,\mathrm{ps}/(\mathrm{mm\,nm})$ and $\Delta f \approx 320\,\mathrm{GHz}$, Fig. 3.9(a). The group index n_g increases linearly with frequency in the region of constant dispersion, and design 1 with larger negative chromatic dispersion has higher group index than design 2, Fig. 3.9(b). Both designs are membrane structures with slab thickness $h = 220\,\mathrm{nm}$. The structure parameters of design 1 are: $W_1 = 0.9\sqrt{3}\,a$, $W_2 = 0.8\tfrac{1}{2}\sqrt{3}\,a$, $W_3 = 0.9\tfrac{1}{2}\sqrt{3}\,a$, $r_1 = r_2 = r = 0.3\,a$, $r_3 = 0.35\,a$, and $a = 460\,\mathrm{nm}$. The parameters of design 2 are $W_1 = 0.9\sqrt{3}\,a$, $W_2 = 0.83\tfrac{1}{2}\sqrt{3}\,a$, $W_3 = 0.87\tfrac{1}{2}\sqrt{3}\,a$, $r_1 = 0.28\,a$, $r_2 = r = 0.3\,a$, $r_3 = 0.35\,a$, and $a = 452\,\mathrm{nm}$.

Constant Positive Chromatic Dispersion

Section 1 of the delay line should exhibit a constant positive chromatic dispersion, with a bandwidth that is larger than that of section 2 having constant negative chromatic dispersion. For the designs of the positive chromatic dispersion section, we start with the basic PC-WGs also used for the negative chromatic dispersion sections, and change the structure parameters such that the chromatic dispersion characteristics is positive in the whole spectral domain; the needed parameter change is rather slight to achieve this. Then, three parts of these PC-WGs whose characteristics are slightly shifted in frequency are concatenated to achieve the constant positive

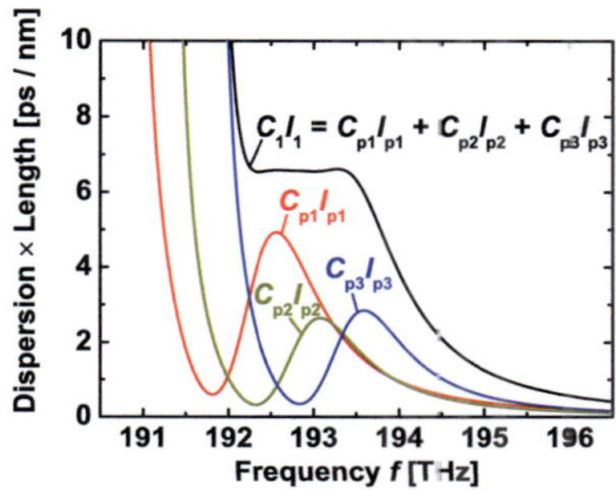

Fig. 3.10. Principle to obtain constant positive chromatic dispersion. A flat dispersion characteristic $C_1 l_1$ can be obtained by concatenating three parts of PC-WG each having positive chromatic dispersion characteristics with slight frequency shifts against each other. The required frequency shifts and lengths of the three parts are found with a parameter fit.

dispersion in a large bandwidth. The resulting dispersion-length product of the section (section 1) $C_1 l_1$ is just the sum of the dispersion-length products of the three parts $C_{p1,2,3} l_{p1,2,3}$; this is expressed by $C_1 l_1 = C_{p1} l_{p1} + C_{p3} l_{p3} + C_{p3} l_{p3}$. The principle is displayed in Fig. 3.10. The shift of the dispersion characteristics in frequency can again be achieved by slightly changing the core width W_1 of the WG. The optimum frequency shifts and lengths of the parts are determined by using a parameter fit. In general, also more than three WG parts can be used, or parameters can be continuously varied over the WG length, which results in a even flatter dispersion characteristics.

In Fig. 3.11, the chromatic dispersion and group index of the two designs are shown, where design 2 is that as already presented previously in Fig. 3.10 with $l_1 = 1\,\text{mm}$. Design 1 has, similarly to the designs for negative chromatic dispersion, a larger positive chromatic dispersion $C \approx 14.6\,\text{ps}/(\text{mm nm})$ and a narrower bandwidth $\Delta f \approx 800\,\text{GHz}$ than design 2 with $C = 6.6\,\text{ps}/(\text{mm nm})$ and $\Delta f \approx 1,200\,\text{GHz}$, Fig. 3.11(a). The group index n_g decreases linearly with frequency in the region of constant dispersion, and design 1 exhibiting larger chromatic dispersion has higher group index than design 2, Fig. 3.9(b). In comparison to the designs for negative chromatic dispersion, only the parameters r_2 and r_3 have been changed. For completeness, we repeat all structure parameters here. The membrane slab thickness is $h = 220\,\text{nm}$. Design 1 has the following parameters: $W_1 = 0.9\sqrt{3}\,a$, $W_2 = 0.8\frac{1}{2}\sqrt{3}\,a$, $W_3 = 0.9\frac{1}{2}\sqrt{3}\,a$, $r_1 = r = 0.3\,a$, $r_2 = 0.28\,a$, $r_3 = 0.34\,a$, and $a = 460\,\text{nm}$. The parameters of design 2 are $W_1 = 0.9\sqrt{3}\,a$, $W_2 = 0.83\frac{1}{2}\sqrt{3}\,a$, $W_3 = 0.87\frac{1}{2}\sqrt{3}\,a$, $r_1 = 0.28\,a$, $r_2 = 0.27\,a$, $r_3 = 0.33\,a$, $r = 0.3\,a$ and $a = 452\,\text{nm}$.

3.2.3 Optimization

For a tunable delay line, the tuning range of the group delay Δt_g is the most relevant parameter, while the bandwidth should be sufficiently large for fast optical signals. Our aim is to obtain a bandwidth of 125 GHz, which allows transmission of at least 40 Gbit/s signals. As mentioned in

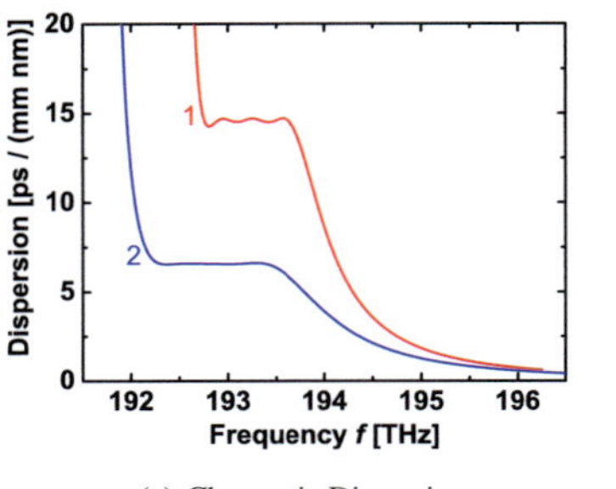

(a) Chromatic Dispersion.

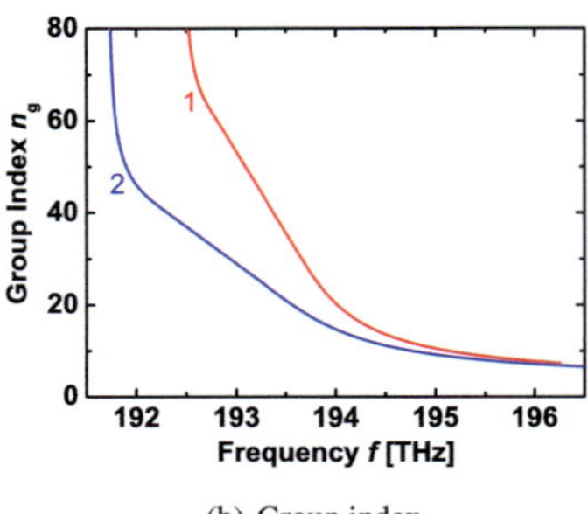

(b) Group index.

Fig. 3.11. PC-WG devices with constant positive chromatic dispersion. (a) Design 1 has a larger positive dispersion $C = 14.6\,\mathrm{ps}/(\mathrm{mm\,nm})$ and a narrower bandwidth $\Delta f = 800\,\mathrm{GHz}$ than design 2 with $C = 6.6\,\mathrm{ps}/(\mathrm{mm\,nm})$ and $\Delta f = 1200\,\mathrm{GHz}$. (b) The group index n_g decreases linearly with frequency in the region of constant dispersion, and the design 1 with larger negative chromatic dispersion has higher group index than design 2. Structure parameters are mentioned in the text.

Subsection 3.2.1, the bandwidth of the suggested device is limited by the bandwidth Δf_2 of section 2 having negative chromatic dispersion. For the two presented designs in Subsection 3.2.2, bandwidths of $\Delta f_2 \approx 200\,\mathrm{GHz}$ or $\Delta f_2 \approx 320\,\mathrm{GHz}$ were estimated, which means we might not need the full available bandwidth.

In the following, we estimate the group delay tuning range Δt_g and the parameters on which it depends to find out how the tuning range can be maximized. According to Eq. (3.6), the group delay variation $\Delta t_{\mathrm{g}1,2}$ of sections 1 or 2 inside the bandwidth $\Delta f_{1,2}$ of constant positive or negative dispersion, respectively, can be calculated as

$$\Delta t_{\mathrm{g}1,2} = \frac{c\,l_{1,2}\,|C_{1,2}|\,\Delta f_{1,2}}{f_0^2}, \tag{3.7}$$

where f_0 is the center frequency and $l_{1,2}$ are the lengths of the sections. According to Fig. 3.7, the achievable group delay difference is

$$\Delta t_\mathrm{g} = \Delta t_{\mathrm{g}1} - \Delta t_{\mathrm{g}2} \tag{3.8}$$

$$= \frac{c}{f_0^2}\left(l_1 C_1 \Delta f_1 - l_2 |C_2| \Delta f_2\right). \tag{3.9}$$

The condition for a resulting flat group delay or zero dispersion is $l_1 C_1 = l_2 |C_2|$, and thus the total length of the device is $L_\mathrm{tot} = l_1 + l_2 = l_1\left(1 + C_1/|C_2|\right)$. Using these two relations, the group delay difference can be expressed as

$$\Delta t_\mathrm{g} = L_\mathrm{tot}\,\frac{c}{f_0^2}\,\frac{C_1\,|C_2|}{C_1 + |C_2|}\left(\Delta f_1 - \Delta f_2\right) \tag{3.10}$$

$$= L_\mathrm{tot}\,\frac{c}{f_0^2}\,\frac{\Delta f_1 - \Delta f_2}{1/C_1 + 1/|C_2|}. \tag{3.11}$$

Clearly, the group delay tuning range can for fixed length L_{tot} be maximized by maximizing both dispersion values C_1 and $|C_2|$, and by maximizing the bandwidth difference $\Delta f_1 - \Delta f_2$. Thus, even if the available bandwidth Δf_2 might be larger than required, we would choose only the minimum needed, in our case $\Delta f_2 = 125\,\text{GHz}$. For the two designs of constant positive and negative dispersion presented above in Subsection 3.2.2, the magnitude of the dispersion values has been maximized. It needs to be mentioned, however, that the magnitude of the chromatic dispersion can in practice not be arbitrarily increased, as the group velocity decreases simultaneously in the PC-WG and losses are thus increased; this might become the major limitation for the device.

For the two presented designs, the group delay difference for devices with a total length of $1\,\text{mm}$ can be calculated, where we fix $\Delta f_1 = 125\,\text{GHz}$. For design 1 with $C_1 = 14.6\,\text{ps}/(\text{mm}\,\text{nm})$, $\Delta f_1 = 800\,\text{GHz}$ and $C_2 = -16.8\,\text{ps}/(\text{mm}\,\text{nm})$, we obtain $\Delta t_g = 42.2\,\text{ps}$. This corresponds to 5.1 pulsewidths for a $40\,\text{Gbit/s}$ 33% RZ data signal. The length ratio of the two sections is $l_1/l_2 = 1.15$. For design 2 with $C_1 = 6.6\,\text{ps}/(\text{mm}\,\text{nm})$, $\Delta f_1 = 1,200\,\text{GHz}$ and $C_2 = -7.9\,\text{ps}/(\text{mm}\,\text{nm})$, the tunable delay is $\Delta t_g = 31.0\,\text{ps}$ corresponding to 3.8 pulsewidths of a $40\,\text{Gbit/s}$ 33% RZ data signal; the length ratio is for that case $l_1/l_2 = 1.20$.

Clearly, if the length of the delay line is increased, also the delay range can be increased. In principle, much higher data rates than $40\,\text{Gbit/s}$ can be transmitted if the optical bandwidth of $125\,\text{GHz}$ is fully exploited or the bandwidth Δf_2 is even further increased to the maximal possible value, or if a modulation format is used that is spectrally more efficient.

3.2.4 Conclusion

We present a tunable delay line with a group delay tuning range of $42.2\,\text{ps}$ in a $125\,\text{GHz}$ optical bandwidth, while the device length is only $1\,\text{mm}$. The delay range corresponds to 5.1 pulsewidths of a $40\,\text{Gbit/s}$ 33% RZ data signal, and even higher bit rates are possible with the given optical bandwidth. The delay line is realized by two concatenated PC-WGs having constant dispersion values of opposite sign.

3.3 Compact High-Speed Electro-Optic Modulator

Fast Mach-Zehnder silicon modulators with low operating voltage fabricated in CMOS technology have the potential to considerably cut costs for high-speed optical transceivers. So far, the fastest broadband light modulation in silicon technology was achieved with free-carrier injection, which allowed modulation up to 30 GHz [133]. On the other hand, much higher modulation frequencies can be achieved with organic materials through the virtually instantaneous electro-optic effect. And indeed, modulation at 165 GHz has been shown in a polymer ring resonator [134], however for a signal bandwidth of a few GHz only. Besides providing a fast electro-optic effect, organic materials have a few more advantages. Poled organic materials can have electro-optic coefficients ranging from moderate $r_{33} = 10\,\text{pm}/\text{V}$ [135] to extremely high values of $r_{33} = 170\,\text{pm}/\text{V}$ [65], [136], thus enabling operation with low drive voltage. Further, the organic materials can be infiltrated into silicon structures [17], [137]. This allows combining the highly nonlinear characteristics of the organic materials with the good high-index guiding properties of silicon-on-insulator devices. To reduce even further the structure size, the drive voltage, and the electrical power dissipation of silicon-organic based modulators, resonant elements [138], [139], [140] or photonic crystals with low group velocity [61] may be employed.

In this section we propose an ultra-compact silicon-based Mach-Zehnder amplitude modulator with a 78 GHz modulation bandwidth and a drive voltage of 1 V, which allows data transmission at 100 Gbit/s. This is achieved by infiltrating an electro-optic organic materials into a slotted photonic crystal waveguide, thereby making use of the fast electro-optic effects of organic materials, the strong field confinement in slotted waveguides and the slow light interaction enhancements provided by the photonic crystal waveguide, where group velocity and dispersion may be controlled [14], Subsection 2.1.3.

This section is organized as follows: In Subsection 3.3.1, we introduce the Mach-Zehnder modulator. Subsection 3.3.2 explains the optimization strategy. Subsection 3.3.3 gives details of the slow-wave phase modulator design. In Subsection 3.3.4, modulator parameters like modulation bandwidth and π-voltage are discussed. Subsection 3.3.5 is devoted to performance data of an optimized integratable Mach-Zehnder modulator, and in Subsection 3.3.6 a taper is proposed for coupling efficiently to the slow-light mode of the phase modulator sections. The Appendix A.7 gives derivations of important relations.

The contents of this section have been published in a journal article, [1] or [J3] on page 165.

3.3.1 The Modulator

The configuration of the integrated Mach-Zehnder (MZ) interferometer (MZI) modulator (MZM) in silicon-on-insulator (SOI) technology is shown schematically in Fig. 3.12. The optical strip waveguides are operated in quasi-TE mode, where the optical field has a dominant electric field component E_x oriented parallel to the substrate plane. Y-branches split and combine the signals. Both arms of the MZ interferometer comprise phase modulator (PM) sections of length L. The PM sections consist of photonic crystal (PC) line defect waveguides (WGs) with a narrow gap of width W_{gap} in the center, and are infiltrated with a $\chi^{(2)}$-nonlinear material (Pockels effect). In such a PC-WG, the dominant optical field component E_x is strongly confined to the gap (see Subsection 3.3.3). The nonlinear material is assumed to be poled [136] such that the axis of strongest electro-optic interaction (arrow ⇑ in Fig. 3.12) is aligned along

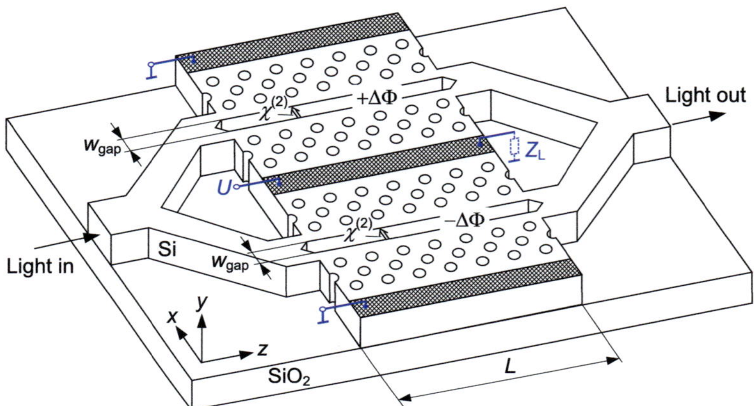

Fig. 3.12. Mach-Zehnder modulator schematic. The input WG carries a quasi-TE mode, the dominant electric field component of which (E_x) is oriented along the x-direction. A Y-branch (in reality an MMI coupler) splits the input into two arms where PC phase modulators are inserted. A coplanar transmission line provides electric bias and a modulation field driving the phase modulators in push-pull mode. The optical signals in both arms experience phase shifts $+\Delta\Phi$ and $-\Delta\Phi$.

the x-direction; the associated electro-optic coefficient is r_{33}. Therefore, only the x-components of the optical field and of the modulating field are relevant. The nonlinear material is assumed to respond instantaneously.

The silicon material of the PM sections is doped, e.g., with arsenic ($n_\mathrm{D} \approx 2 \times 10^{16}\,\mathrm{cm}^{-3}$), to be sufficiently conductive ($\sigma = 10\,\Omega^{-1}\mathrm{cm}^{-1}$) without introducing excessive optical loss. The edges of the PM silicon slabs are metalized with aluminum on top, and the three metallic layers (dark shading) running in parallel to the nonlinear optical WGs form a microwave coplanar waveguide (CPWG). The two outer electrodes are grounded, and a modulating voltage U applied to the center electrode generates a voltage wave that travels along the line and drives the PM sections in push-pull mode. In each of the arms, the electric modulation field $E_\mathrm{el} = U/W_\mathrm{gap}$ is dominantly oriented along the x-direction and almost completely confined to the gap. An optical quasi-TE field is launched at the input side of the MZM in Fig. 3.12. The phase shifts induced on the two MZI arms by the respective PMs are $+\Delta\Phi = +[\Delta\Phi_0 + \Delta\Phi(t)]$ and $-\Delta\Phi = -[\Delta\Phi_0 + \Delta\Phi(t)]$, where $\Delta\Phi(t) \sim U(t)$ represents the time-varying part. The bias voltage U_0 is chosen such that the offset point is set to $\Delta\Phi_0 = \pi/4$ (total phase difference $2\Delta\Phi_0 = \pi/2$ between both arms, MZ interferometer in quadrature). For switching the MZM from full transmission to full extinction, the phase in each PM section must be changed from $\Delta\Phi = -\pi/4$ to $\Delta\Phi = +\pi/4$, which corresponds to a voltage change from $U = -U_\pi/4$ to $U = +U_\pi/4$. The modulation voltage U that is required to change the phase in one modulator arm by π is called π-voltage U_π.

If the modulator is realized as shown in Fig. 3.12, the Y-junctions and the transitions from strip-WG to slow-light slot PC-WG lead to significant reflection and light scattering, and the minimum feature sizes cannot be fabricated accurately. For a more practical design, the Y-branches should be replaced by MMI couplers. Further, an efficient transition from strip-WG to slow-light slot PC-WG is used as explained in Subsection 3.3.6.

3.3.2 Mach-Zehnder Optimization Strategy

An optimum MZM should provide large modulation bandwidth $f_{3\,\mathrm{dB}}$, require low modulation voltage amplitude U, should be fabricated based on CMOS processes, and it should have small size for easy on-chip integration. The MZM modulation bandwidth is in general limited by RC-effects and by the spatial walk-off between the electrical and optical waves. The spatial walk-off can be avoided by a traveling-wave design, where the group velocities $v_{\mathrm{g,el}}$ and $v_{\mathrm{g,opt}}$ of the electrical and optical waves are same, and where the CPWG is terminated with a matching impedance Z_{L} to avoid reflections.

However, such a traveling-wave structure with a well-terminated CPWG is difficult to realize, especially in a wide optical and electrical frequency range. If the modulator can be made sufficiently short in length L, a match of $v_{\mathrm{g,el}}$ and $v_{\mathrm{g,opt}}$ is not needed. We achieve this length reduction by using a PC line defect WG that is designed to have low optical group velocity and thus shows an increased electro-optic interaction. The maximum modulation frequency $f_{3\,\mathrm{dB}}$ is reached when the electrical modulation signal inside the electrical structure varies by more than half a period during the propagation time $t_{\mathrm{g,opt}} = v_{\mathrm{g,opt}}/L$ of the slow optical wave through the PM, see left-hand side of Eq. (3.12). This maximum frequency corresponds to the walk-off related bandwidth derived in Subsection 3.3.4 under the condition $v_{\mathrm{g,opt}} \ll v_{\mathrm{g,el}}$. In addition, we find that RC-effects do not play a significant role for our structure (see Appendix A.7). Further, we derive in Subsection 3.3.4 and in the Appendix A.7 a relation for the π-voltage U_{π}, which in a simplified form is given by the right-hand side of Eq. (3.12),

$$ f_{3\,\mathrm{dB}} \approx \frac{0.5}{t_{\mathrm{g,opt}}} = \frac{0.5\,v_{\mathrm{g,opt}}}{L}, \qquad U_{\pi} \propto \frac{W_{\mathrm{gap}}}{r_{33}}\frac{v_{\mathrm{g,opt}}}{L}. \tag{3.12} $$

From Eq. (3.12) we observe that the modulation bandwidth increases with the ratio $v_{\mathrm{g,opt}}/L$, however, at the expense of an increased π-voltage, i.e., a large drive voltage amplitude. For an optimum modulator (large bandwidth, low drive voltage) we need fixing the design bandwidth $f_{3\,\mathrm{dB}}$ first, i. e., we have to decide for a certain ratio $v_{\mathrm{g,opt}}/L$. The modulator length L is then directly related to the optical group velocity $v_{\mathrm{g,opt}}$, and a lower group velocity results in a shorter length. To reduce the π-voltage, a PM material with large nonlinear coefficient r_{33} needs to be chosen, and the gap width W_{gap} of the slotted waveguide should be made small. As we show in Subsection 3.3.3, both the optical and the microwave fields remain strongly confined to the nonlinear material, even for a narrow gap width W_{gap}. This is a specific advantage of the slotted PC-WG.

3.3.3 Slow-Wave Phase Modulator

Figure 3.13 displays the slow-wave phase modulator section in more detail. In the center of a PC line defect WG, a narrow gap W_{gap} is cut out in form of a slot. The PC consists of a silicon slab with a triangular lattice of air holes having a lattice constant a. For a Wi line defect, a number of i rows of holes are omitted. The width W_1 of the resulting waveguide, see Fig. 3.13(b), depends on i, which need not be an integer. As explained previously, the silicon structure is covered with a highly nonlinear poled electro-optic organic material, which fills both the slot and the PC holes. Due to the high index-contrast between silicon ($n_{Si} = 3.48$) and the organic electro-optic material ($n_{EO} = 1.6$), the optical quasi-TE mode is mainly confined to the gap filled with the organic material, Fig. 3.13(b) and [98].

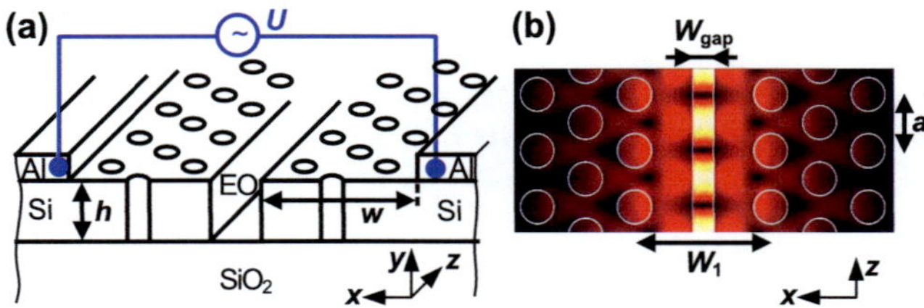

Fig. 3.13. Phase modulator (a) schematic and (b) dominant electric field component E_x. A slot filled with an electro-optic organic material (EO) of width W_{gap} is cut in a silicon photonic crystal line-defect waveguide of width W_1. The silicon slabs of height h and width w are doped for electrical conductivity and contacted with aluminum layers. E_x is strongly confined to the slot. The phase $\Delta\Phi$ of the propagating optical wave is tuned by applying a voltage to the organic material. The triangular-lattice period is a.

For a minimum modulation voltage U we need to avoid any voltage drop in the silicon material between the electrodes (Al in Fig. 3.13(a)) and the gap. To this end, the silicon must be made sufficiently conductive by doping. Choosing then the smallest possible gap that is still compatible with technological constraints, and selecting an organic material with a large linear electro-optic coefficient minimizes the π-voltage U_π, Eq. (3.12).

While maintaining the low group velocity of the optical mode, the PC-WG can be adjusted such that chromatic dispersion of the optical wave is negligible, and signal distortions are avoided. In Subsection 3.3.5 it is shown that such a design leads to a large MZM bandwidth $f_{3dB} = 78\,\text{GHz}$ for a drive voltage amplitude as small as $\hat{U} = U_\pi/4 = 1\,\text{V}$. For explaining our PC waveguide design decisions, we first describe a slow-wave PC slot waveguide PM and its properties, and then discuss the optimized PC slot waveguide design for a flattened dispersion curve.

PC Slot Waveguide

PC waveguides lend themselves easily to a reduced group velocity near the edge of the Brillouin zone. A conventionally designed PC slot WG usually supports a mode with low group velocity, indeed. For a line-defect width W_1 chosen to correspond to a W1.4 line-defect WG, the associated band diagram is displayed in Fig. 3.14. The low group velocity region is marked by an oval to the right. The figure is calculated with the guided-mode expansion (GME) method [141], and simulations with the finite integration technique (FIT) verify the results.

The group velocity of the mode inside the marked region of Fig. 3.14 is displayed in Fig. 3.15(b) as a function of frequency (dashed line). The group velocity varies between nearly 0 % and 6 % of the vacuum speed of light c in the region below the light line. However, a high chromatic dispersion of $C > 5\,\text{ps}/(\text{mm}\,\text{nm})$ is also observed.

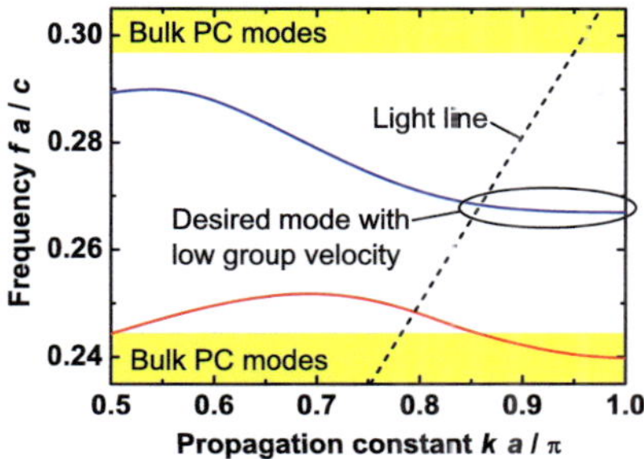

Fig. 3.14. Band diagram of W1.4 PC slot waveguide. The desired mode exhibits a low group velocity below the light line of the organic cladding. PC slab height $h = 220\,\text{nm}$, organic material gap width $W_{\text{gap}} = 150\,\text{nm}$, PC lattice period $a = 408\,\text{nm}$, hole radii $r/a = 0.3$, line defect width $W_1 = 1.4\sqrt{3}\,a$. The refractive index of the organic electro-optic material is $n_{\text{EO}} = 1.6$, f, k, c denote frequency, propagation constant and vacuum speed of light, respectively.

PC Slot Waveguide with Dispersion Engineering

For a better design with a lower chromatic dispersion, we optimize the hole radii $r_{1,2,3}$ and the distances between the hole centers $W_{1,2,3}$, see Fig. 3.15(a). As a result of the optimization process we find a set of W1.25 WGs that all provide a low group velocity over a wide spectral range. The frequency dependence of the resulting group velocity with various radii r_2 as parameters is shown in Fig. 3.15(b). An increase of the parameter r_2 decreases the group velocity, while a flat dispersion is maintained. For the value $r_2/a = 0.36$, the group velocity is 4% of the vacuum speed of light c over a bandwidth of about $1\,\text{THz}$. It is also possible to obtain a negative chromatic dispersion, which is for example the case for $r_2/a = 0.30$ or $r_2/a = 0.38$, Fig. 3.15(b). For all presented designs, the air hole diameters are larger than $200\,\text{nm}$ in order to meet fabrication constraints. The concept of the PC-WG with broadband low group velocity is experimentally verified in Section 4.1.

3.3.4 Modulator Performance Parameters

The main performance parameters for the modulator are the MZM modulation bandwidth $f_{3\,\text{dB}}$ and the phase modulator π-voltage U_π. The two parameters are derived and discussed in this subsection.

Modulation Bandwidth of Mach-Zehnder Modulator

The bandwidth of the MZM is limited by the walk-off between electrical and optical waves. *RC* limitations do not play a role for the presented structures as has been shown in Appendix A.7. The walk-off limited bandwidth depends on the termination of the CPWG, and here we discuss the two cases of a matched load and an open.

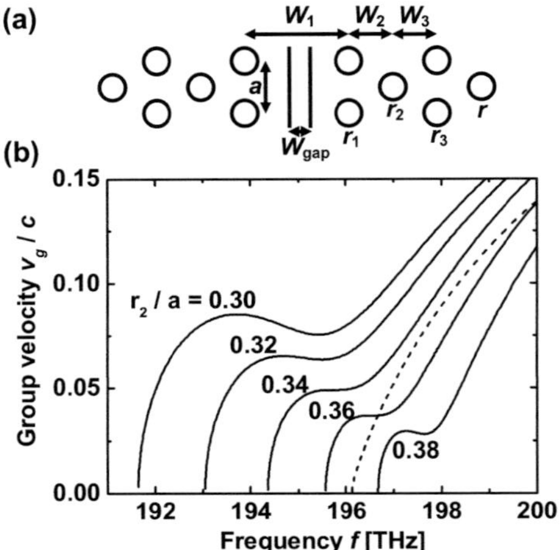

Fig. 3.15. W1.25 PC slot waveguide with slab height $h = 220\,$nm, polymer gap width $W_{\mathrm{gap}} = 150\,$nm, PC lattice period $a = 408\,$nm. (a) Structure parameters and (b) group velocity as a function of frequency with varying hole radii r_2. With $r_2/a = 0.36$, the group velocity amounts to 4 % of the vacuum speed of light c, and the group velocity dispersion is negligible in a bandwidth of 1 THz. For comparison, the group velocity of the conventional W1.4-WG of Fig. 2 is plotted as a dashed line ($W_1 = 1.4\sqrt{3}\,a$, $W_2 = W_3 = 0.5\sqrt{3}\,a$, $r_1 = r_2 = r_3 = r = 0.3\,a$). Parameters of the W1.25-WG are $W_1 = 1.25\sqrt{3}\,a$, $W_2 = 0.65\sqrt{3}\,a$, $W_3 = 0.45\sqrt{3}\,a$, $r_1 = 0.25\,a$, $r_3 = r = 0.3\,a$.

1) Walk-off bandwidth, CPWG with matched load

The CPWG be ideally terminated with a matched load so that the modulating wave traveling along z is not reflected at the end of the CPWG and maintains a spatially constant amplitude $|U|$. Electrical and optical waves propagate in the same direction, but in general at different group velocities $v_{\mathrm{g,el}}$ and $v_{\mathrm{g,opt}}$ (group delays $t_{\mathrm{g,el}}$ and $t_{\mathrm{g,opt}}$ over the PM section length L). The nonlinear interaction is maximum for co-directionally traveling waves (TW) with $t_{\mathrm{g,el}} = t_{\mathrm{g,opt}}$, and it decreases strongly if $t_{\mathrm{g,el}} \neq t_{\mathrm{g,opt}}$. When the electrical and optical signal envelopes have acquired a phase difference of π, the limiting modulating frequency $f_{3\mathrm{dB}}$ is reached with $\omega_{3\,\mathrm{dB}}^{(\mathrm{TW})} |t_{\mathrm{g,opt}} - t_{\mathrm{g,el}}| = \pi$, i. e.,

$$f_{3\,\mathrm{dB}}^{(\mathrm{TW})} = \frac{0.5}{|t_{\mathrm{g,opt}} - t_{\mathrm{g,el}}|} = \frac{0.5\,v_{\mathrm{g,opt}}}{L}\frac{1}{|1 - v_{\mathrm{g,opt}}/v_{\mathrm{g,el}}|}. \tag{3.13}$$

Following the formalism in [142], a more accurate formula can be derived, see Appendix A.7, whereby it is shown that the factor of 0.5 in Eq. (3.13) need be replaced by 0.556.

2) Walk-off bandwidth, CPWG with open-circuit

In order to keep the design of an integratable MZM simple, the CPWG would be more favorably be configured without terminating resistances. Because of the reflection at the open CPWG, a standing wave results. For the forward propagating part of the electrical wave, Eq. (3.13) specifies the limiting frequency. However, the backward propagating wave puts a tighter walk-off limit, because now the electrical wave has the opposite direction compared to the optical wave. With similar arguments that led to Eq. (3.13) we find $\omega_{3\,\mathrm{dB}}|t_{\mathrm{g,\,opt}}+t_{\mathrm{g,\,el}}| = \pi$,

$$f_{3\,\mathrm{dB}} = \frac{0.5}{|t_{\mathrm{g,\,opt}}+t_{\mathrm{g,\,el}}|} = \frac{0.5\,v_{\mathrm{g,\,opt}}}{L}\,\frac{1}{|1+v_{\mathrm{g,\,opt}}/v_{\mathrm{g,\,el}}|}. \tag{3.14}$$

As above, a more accurate formula can be developed following [142] where the factor of 0.5 in Eq. (3.14) is again replaced by 0.556. If $v_{\mathrm{g,\,opt}} \ll v_{\mathrm{g,\,el}}$ holds, then $f_{3\,\mathrm{dB}} \approx f_{3\,\mathrm{dB}}^{(\mathrm{TW})}$ and electrically short MZM designs with and without matching load for the CPWG become nearly equivalent, resulting in Eq. (3.12).

π-Voltage U_π of Phase Modulator

A Mach-Zehnder modulator's sensitivity is characterized by the π-Voltage U_π of its phase modulators. For a given PM length L the voltage $|U|$ needed for a π-phase shift is defined to be U_π. For large modulation sensitivity, U_π should be small. An optical wave propagating through a PM experiences a nonlinear refractive index change Δn in proportion to the electric modulating field E_{el} inside the nonlinear PM material,

$$\Delta n = -\frac{1}{2} r_{33}\, n_{\mathrm{EO}}^3\, E_{\mathrm{el}}, \qquad E_{\mathrm{el}} = U/W_{\mathrm{gap}}. \tag{3.15}$$

As before, n_{EO} represents the effective linear part of the refractive index in the PM section, and r_{33} is the (scalar) linear electro-optic coefficient. The total phase shift of the optical wave due to the index change in a PM section of length L is (following the formalism in Appendix A.7 and in [142])

$$\Delta\Phi = -\Delta\beta L = -\Gamma \Delta n k_0 L. \tag{3.16}$$

The quantity $k_0 = 2\pi f_0/c$ is the vacuum wave number. In the process of deriving Eq. (3.16) the so-called field interaction factor Γ was introduced. It quantifies the strength of the nonlinear electro-optic interaction of modulating field and optical mode in a cross-section A along a lattice period a (see Appendix Eqs. (A.76)-(A.80) and [142]),

$$\Gamma = \int_{\mathrm{gap}} \frac{n_{\mathrm{EO}}}{Z_0} |\hat{E}_x|^2 \,\mathrm{d}V \Big/ \int a\,\Re(\hat{\mathbf{E}} \times \hat{\mathbf{H}}^*) \cdot \mathbf{e}_z \,\mathrm{d}A \propto \frac{1}{v_{\mathrm{g,\,opt}}}. \tag{3.17}$$

In Eq. (3.17), Z_0 is the free-space wave impedance, $\hat{\mathbf{E}}$ (x-component E_x) and $\hat{\mathbf{H}}$ are the optical modal electric and magnetic fields, and $\mathbf{e}_z$ is the unit vector in z-direction. The field *interaction* factor Γ as defined in Eq. (3.17) is different from the field *confinement* factor, which is usually calculated as a ratio between the optical power in the cross-section of the interaction region and the total power propagating in the whole modal cross-section. While the field confinement factor varies between 0 and 1, the field interaction factor Γ can be larger than 1. The integral in the numerator of Eq. (3.17) is a measure of the energy stored in the transverse component of the propagating optical mode inside the PM along a length a (see Appendix, Eq. (A.18)). The denominator represents the transported power in the modal cross-section. According to Eq. (A.19), the energy stored in a volume, when related to the cross-section power of a wave that crosses this volume, increases in proportion to the reciprocal group velocity of the wave. Because in the numerator of Γ only the dominant transverse field component is regarded, the proportionality $\Gamma \propto 1/v_{g,\,\mathrm{opt}}$ holds only approximately. Application of a voltage U_π to the PM results by definition in a phase change of $\Delta\Phi = \pi$ within a length L. From Eq. (3.15)-(3.17), the π-voltage is computed to be

$$U_\pi = \frac{c}{n_{\mathrm{EO}}^3 f_0} \frac{W_{\mathrm{gap}}}{r_{33}} \frac{1}{L\Gamma}, \qquad U_\pi \propto \frac{W_{\mathrm{gap}}}{r_{33}} \frac{1}{L\Gamma} \propto \frac{W_{\mathrm{gap}}}{r_{33}} \frac{v_{g,\,\mathrm{opt}}}{L} \propto \frac{W_{\mathrm{gap}}}{r_{33}} f_{3\,\mathrm{dB}}. \tag{3.18}$$

For the proportionality at the right-hand side of Eq. (3.18), the relations $\Gamma \propto 1/v_{g,\,\mathrm{opt}}$ and $f_{3\,\mathrm{dB}} \propto v_{g,\,\mathrm{opt}}/L$ were substituted from Eq. (3.17) and Eq. (3.14).

3.3.5 Optimized Mach-Zehnder Modulator

For maximizing the modulation bandwidth $f_{3\,\mathrm{dB}}$, Eq. (3.14), of the MZM amplitude modulator and for minimizing its π-voltage U_π, Eq. (3.18), the optical group velocity $v_{g,\,\mathrm{opt}}$ of the PC line defect WG, its length L, the electro-optic coefficient r_{33} of the organic material, and the gap width W_{gap} need to be adjusted properly. For an integratable MZM with small length L, the design bandwidth $f_{3\,\mathrm{dB}}$ fixes the ratio $v_{g,\,\mathrm{opt}}/L$. Reducing $v_{g,\,\mathrm{opt}}$ then provides a small length L. It also needs to be considered that with lower $v_{g,\,\mathrm{opt}}$, the disorder-induced losses of the PC-WG increase [58] and the optical bandwidth decreases, so that $v_{g,\,\mathrm{opt}}$ cannot be made arbitrarily small.

Structure	r_2/a	f_0 (THz)	$v_{g,\mathrm{opt}}/c$	Γ	L (μm)	$f_{3\,\mathrm{dB}}$ (GHz)
	0.3	196.4	2.4%	4.8	36	103
W1.4		196.6	3.4%	3.2	54	97
dispersion		196.9	4.8%	2.2	80	90
large		196.2	6.4%	1.5	113	83
		197.7	8.2%	1.1	155	76
	0.38	197.5	3.2%	3.1	57	87
W1.25	**0.36**	**196.5**	**4.0%**	**2.2**	**80**	**78**
dispersion	0.34	195.8	5.2%	1.6	111	71
flattened	0.32	195.2	6.6%	1.1	158	61
	0.30	194.6	7.9%	0.8	215	53

Table 3.1. Characteristic data for a PC slot waveguide modulator. Group velocity $v_{g,\mathrm{opt}}$, field interaction factor Γ, modulator length L and modulation bandwidth $f_{3\,\mathrm{dB}}$ are estimated at different optical carrier frequencies f_0. We assume an electro-optic coefficient of $r_{33} = 80\,\mathrm{pm/V}$. The modulation voltage amplitude for maximum extinction is fixed to $\hat{U} = U_\pi/4 = 1\,\mathrm{V}$.

For a small U_π, the electro-optic organic material is chosen to have a large linear electro-optic coefficient of $r_{33} = 80\,\mathrm{pm/V}$ [17]. Further, W_{gap} is chosen as small as compatible with the fabrication process. A gap width of $W_{\mathrm{gap}} = 150\,\mathrm{nm}$ can be fabricated to good accuracy with advanced lithographic processes; hence this width is fixed for the present design. Given the gap width W_{gap}, the maximum modulation voltage amplitude is limited in practice by the microwave source and by dielectric breakdown in the gap.

In Table 3.1, we list group velocity $v_{\mathrm{g,opt}}$, field interaction factor Γ, modulator length L, and modulation bandwidth $f_{3\,\mathrm{dB}}$ for various PC slot waveguide modulators without (W1.4) and with dispersion flattening (W1.25), as already presented in Subsection 3.3.3. The π-voltage was kept fixed to $U_\pi = 4\,\mathrm{V}$ in all cases, which means that the modulation voltage amplitude $\hat{U} = U_\pi/4 = 1\,\mathrm{V}$ remained constant by adjusting the length L according to Eq. (3.18).

The values $v_{\mathrm{g,opt}}$ and Γ are calculated from simulations with the FIT method. As expected, the field interaction factor Γ increases and the modulator length L decreases when lowering the group velocity $v_{\mathrm{g,opt}}$ according to Eqs. (3.17), (3.18). The bandwidth $f_{3\,\mathrm{dB}}$ is calculated from Eq. (3.14) using the more exact numerical factor of 0.556 instead of 0.5. For a constant U_π the estimate Eq. (3.12) would predict a constant modulation bandwidth $f_{3\,\mathrm{dB}}$, however, it shows a weak $(1:1.4)$ dependence on $v_{\mathrm{g,opt}}$ $(1:3.4)$. This is explained after Eq. (3.17): The field interaction factor Γ is only approximately proportional to $1/v_{\mathrm{g,opt}}$, and therefore $U_\pi \propto v_{\mathrm{g,opt}}$ in Eqs. (3.18), (3.12) is an approximation, too. For the dispersion flattened structures, $f_{3\,\mathrm{dB}}$ is lower and shows a stronger dependence on $v_{\mathrm{g,opt}}$ compared to the structure with high dispersion.

The disorder-induced losses of a slow-light PC-WG operated at a group velocity of 5.8 % of the vacuum speed of light were measured to be $4.2\,\mathrm{dB}/\mathrm{mm}$ [93]. Doping silicon with a large concentration of phosphorus atoms $(2 \times 10^{18}\,\mathrm{cm}^{-3})$ leads to additional optical losses of only about $1\,\mathrm{dB}/\mathrm{mm}$ [137]. At a slightly smaller group velocity of 4 % of the vacuum speed of light and a moderate doping concentration of $2 \times 10^{16}\,\mathrm{cm}^{-3}$, optical losses will be mainly caused by disorder. With small device lengths the additional loss is expected to be tolerable.

Because the dispersion-flattened structure shows a large modulation bandwidth of $f_{3\,\mathrm{dB}} = 78\,\mathrm{GHz}$ at a length of only $L = 80\,\mu\mathrm{m}$ and for a small $1\,\mathrm{V}$ drive voltage amplitude, we regard this to be an optimum structure for the discussed technological constraints.

3.3.6 Slow-Light Coupling Structure

Signals from an external fiber may be effectively coupled to a conventional strip WG mode with coupling losses below $1\,\mathrm{dB}$ [34]. However, an efficient method is also needed to excite

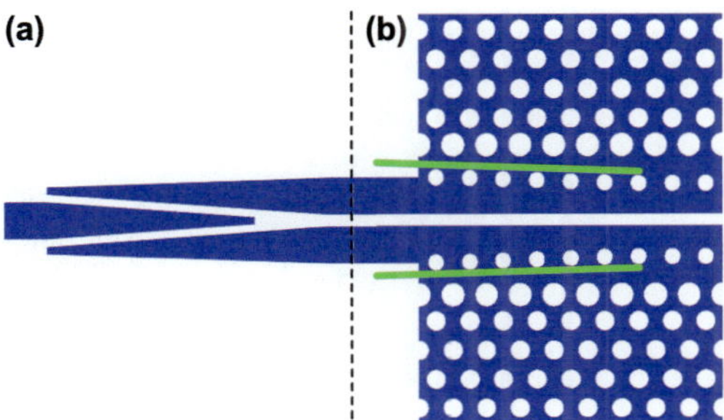

Fig. 3.16. Schematic of the coupling structure. (a) Transition from strip-WG to slot-WG (b) Coupling to PC-WG. The transmission is significantly increased by introducing a PC taper, where the width W_1 of the PC-WG is slightly decreased from $1.45\sqrt{3}\,a$ to $1.25\sqrt{3}\,a$ over some lattice periods, indicated by the overlaid tilted (green) lines, and the width W_2 is increased from $0.55\sqrt{3}\,a$ to $0.65\sqrt{3}\,a$. The width of the strip-WG is 440 nm, and the gap width of both the slot-WG and the PC-WG is 150 nm.

the slow-light mode within the PC slot WG. We propose a coupling structure consisting of two sections, which are schematically shown in Fig. 3.16.

The first section transforms the strip WG mode into a slot-WG mode, Fig. 3.16(a). The design takes into account that the silicon on both sides of the gap needs to be electrically isolated as it is conductive and carries the modulation voltage, Fig. 3.13(a). FIT simulations of the strip-to-slot WG transition having a length of $7\,\mu$m predict a transmission loss lower than 0.3 dB and a reflection lower than -28 dB.

The second section couples the slot WG to the slow-light PC-WG, Fig. 3.16(b). We developed a taper where the width of the PC-WG is slightly decreased over a length of 10 periods ($4.08\,\mu$m) to gradually slow down the PC-WG mode according to the principles developed in Subsection 2.1.4. The calculated transmission and reflection curves are displayed in Fig. 3.17. The simulated structure comprises both the transition from a slot WG to a slow-light PC-WG, Fig. 3.16(b), and the transition back to a slot WG.

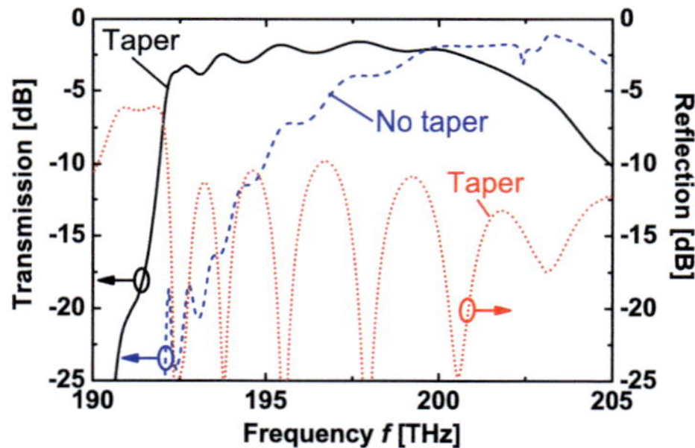

Fig. 3.17. Transmission and reflection for the transition from slot-WG to PC-WG and back to slot-WG, Fig. 5(b), with and without a PC taper. The introduction of the PC taper significantly enhances the transmission to a value better than -4 dB. The reflection is below -10 dB.

The transmission is better than -4 dB including both tapers, while it drops below -20 dB without tapers. The reflection stays below -10 dB. In the transmission curve with tapers, ripples can be observed. These are Fabry-Perot fringes generated by residual reflections at the interfaces. These reflections can be decreased by optimizing the transitions, and an extension of the taper sections lengths can further improve the device characteristics.

3.3.7   Conclusion

We propose a high-speed silicon modulator with low drive voltage based on a slow-light PC-WG. A highly nonlinear organic material is infiltrated into a narrow slot in the WG center, where both the optical and the electric fields are strongly confined. For a design with negligible first-order chromatic dispersion in an optical bandwidth of 1 THz, we predict a modulation bandwidth of 78 GHz and a length of about $80\,\mu$m at a drive voltage amplitude of 1 V. This allows transmission at 100 Gbit/s. We estimate that the modulation bandwidth is limited by the walk-off between the optical and electrical signals, while $RC-$effects do not play a role.

Chapter 4

Experiments and Modeling

In the previous chapters, novel PC-WGs and functional PC devices have been proposed. The presented results are based on various numerical simulation tools, which allow to design the dispersive properties of periodic PC-WGs and to predict the transmission, reflection and group delay characteristics of complete PC devices. The final goal is to fabricate these devices, which first requires that the simulation tools predict the device behavior correctly, and second that the structures can actually be fabricated with current fabrication processes.

In this chapter, we present successful experimental results on the basic PC-WGs needed for the devices, and we test the used simulation tools. We show that the structures are producible with a deep-UV lithographic process suited for mass-production, and that the WGs exhibit the general characteristics as designed. However, in these first fabrication runs, the fabrication-induced imperfections are significant and the device performance is degraded. To this end, we develop in addition an upscaled microwave model which allows to very accurately characterize integrated optical devices and to ultimately verify the design and simulation tools. Out of the various numerical tools, not all are equally efficient and accurate. Thus, we also study and compare the validity and suitability of different simulation tools.

This chapter is organized as follows: In Section 4.1, the optical experiments are presented with results from strip-WGs, dispersion-engineered PC-WGs and slot PC-WGs. Section 4.2 introduces the microwave model and shows the experimental verification of a slow-light PC-WG. Section 4.3 is devoted to the verification and comparison of several simulation tools.

4.1 Optical Experiments

Over the last couple of years, much effort has been made to successfully improve the quality and accuracy of silicon fabrication processes. Especially the fabrication of small optical strip waveguides and photonic crystal structures push lithographic processes at their limit. The two main techniques to obtain accuracies below 100 nm are e-beam lithography [143] and deep-ultraviolet (UV) lithography [144]. While e-beam lithography is very accurate and well suited for research purposes, only devices with small size can be realized. For mass production of devices, deep UV lithography is a feasible method, however to a smaller fabrication accuracy as compared to e-beam lithography. Because of the better fabrication accuracy, e-beam lithography is the more commonly used approach to demonstrate small silicon devices. In recent works, broadband slow-light PC-WGs [93], [94], [95], [96] and PC slot WGs [145] based on the SOI material system have been experimentally demonstrated.

In this work, we have in particular studied broadband dispersion-tailored PC-WG devices. The basic components of these devices are specifically designed PC-WGs and PC slot WGs, and in this section we show experimental results of a broadband slow-light WG, of a WG with linearly varying chromatic dispersion and a negative dispersion minimum, and of a slow-light slot WG. Our structures were fabricated with deep-UV lithography at IMEC (Interuniversity Microelectronics Center) in Leuven, Belgium. The fabrication process uses 193 nm deep-UV lithography and 200 mm silicon-on-insulator (SOI) wafers having a 220 nm top silicon layer and a $2\,\mu$m oxide layer beneath; with such a process, highly-performing strip WGs and PC-WGs have been previously fabricated [146]. To obtain a vertically symmetric structure, we deposit an additional layer of oxide on top of the silicon device layer; we refer to this as a 'buried' structure. For the different fabricated structures, we show scanning electron microscope (SEM) photographs, and measure the transmission and group delay characteristics. The measured group delay is compared to that predicted by simulations, and good agreement is found. We demonstrate a PC-WG with broadband slow light propagation at 3.7 % of the vacuum speed of light with losses of 14.0 dB / mm. The PC-WG with linearly varying chromatic dispersion is shown to exhibit a negative dispersion minimum of $-5\,$ps / (mm nm) and losses of 11.15 dB / mm. The slow-light slot PC-WG has a varying group velocity between 0 % and 12.3 % of the vacuum speed of light and minimum losses of 10 dB / mm.

This section is structured as follows: In Subsection 4.1.1, we introduce the measurement setup, which is followed by measurement results of strip WGs in Subsection 4.1.2, of broadband slow-light WGs in Subsection 4.1.3, of WGs exhibiting linearly varying chromatic dispersion in Subsection 4.1.4, and of slow-light slot WGs in Subsection 4.1.5.

4.1.1　Measurement Setups

For the characterization of the samples, we use two different measurement setups, which are schematically shown in Fig. 4.1. The first setup Fig. 4.1(a) is used to determine the transmission properties of the devices, and the second setup Fig. 4.1(b) is for obtaining the group delay characteristics. The silicon devices are cleaved at the input and output sides, and polarization

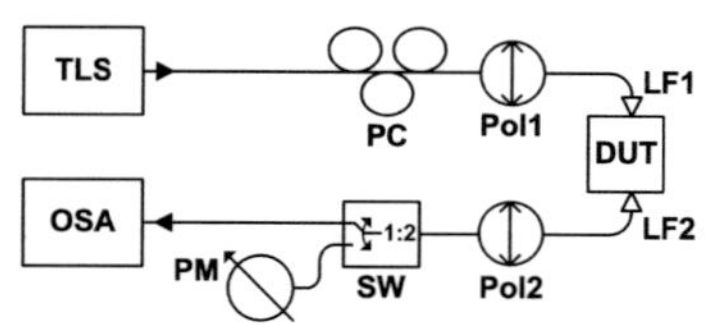

(a) Transmission measurement setup.

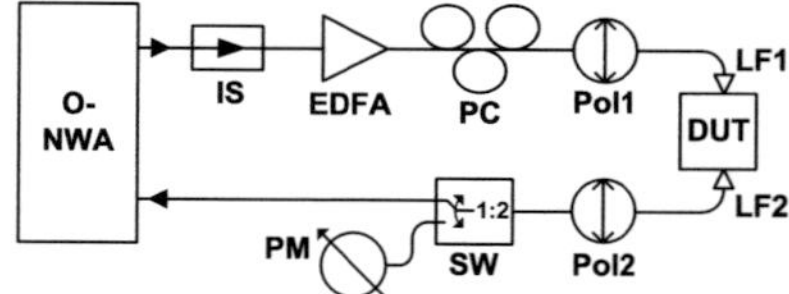

(b) Group delay measurement setup.

Fig. 4.1. Measurement setups for optical device characterization. The silicon device under test (DUT) is coupled in with polarization maintaining lensed fibers (LF). TLS: Tunable laser source. PC: Polarization controller. Pol: Polarizer. SW: Optical switch. PM: Optical power monitor. OSA: Optical spectrum analyzer. O-NWA: Optical network analyzer. IS: isolator. EDFA: Erbium-doped fiber amplifier.

maintaining lensed fibers are used to couple to the silicon WGs at the facets. The fibers can be very accurately positioned via translation stages and piezo controllers. To increase the coupling efficiency from the fibers having low index contrast and large core diameters of $3\,\mu$m to the silicon WGs with high index contrast and a small slab height of 220 nm, the silicon strip-WGs are broadened to a width of $3.24\,\mu$m at the facets. Inside the silicon devices, the strip-WGs are tapered down to the standard width of $440\,\mu$m with single-moded behavior, which then connect the PC-WGs. Because of the cleaved facets of the silicon devices, reflections occur which lead to a resonator-like behavior and result in oscillations in the measured transmission and group delay characteristics.

The transmission setup Fig. 4.1(a) consists of a tunable laser source (Ando AQ4321, 1480 nm − 1580 nm) and an optical spectrum analyzer (Ando AQ6317B), which are simultaneously tuned in frequency. The tunable laser source is connected to a polarization controller, which is followed by a polarizer to select the quasi-TE polarization in the lensed fiber at the input side of the silicon device. At the output side, a second polarizer is included to only consider TE output and to disregard scattering into the TM polarization. An optical switch (EXFO IQ-900) follows, which allows to either connect to a power monitor (Agilent 81531A) for optimizing the coupling between fiber and WG or to the optical spectrum analyzer for the transmission measurement as a function of frequency. The coupling is optimized by measuring the transmitted power at a fixed frequency inside the transmission band of the WG and adjusting the position of the two fibers, which requires an accurate pre-adjustment of the fibers with the help of an optical microscope.

The group delay setup Fig. 4.1(b) is very similar to the transmission setup. The tunable laser source and optical network analyzer are replaced by an optical network analyzer (Advantest Q7750, 1525 nm − 1635 nm) which combines the laser source and the receiver. After the source, an optical isolator is inserted to avoid backreflections from the following high-power amplifier, which is needed to increase the optical power that reaches the receiver. The amplifier is an erbium-doped fiber amplifier (EDFA) from AFC with a maximum output power of 18 dBm. The coupling from the lensed fibers to the silicon WGs is again optimized using the tunable laser source and the power meter, which is more stable than using the optical network analyzer. The group delay is determined by the optical network analyzer with a phase shift method, where the optical carrier is modulated at 3 GHz and the phase shift of the modulation signal is detected as a function of the carrier wavelength. To decrease the drift in the group delay measurements, a reference measurement is performed directly after each device measurement. For the reference measurement, the two fiber tips are directly aligned without any device in-between. The final group delay result is obtained by subtracting the reference delay from the device delay. An unwanted drift in the group delay measurements in the order of $1 - 3$ ps is present, which we attribute to the thermal drift of the fiber amplifier and of the silicon device.

4.1.2 Strip Waveguide

We start with the characterization of a conventional strip-WG with a nominal width of 440 nm. A SEM photograph of a section of the WG is shown in Fig. 4.2. At the sidewalls, surface roughness can be seen, which is caused by fabrication imperfections due to the limited resolution of the lithographic process; this surface roughness is the main source of transmission losses. Furthermore, white shadings at the boundaries of the WG can be seen which indicate non-vertical

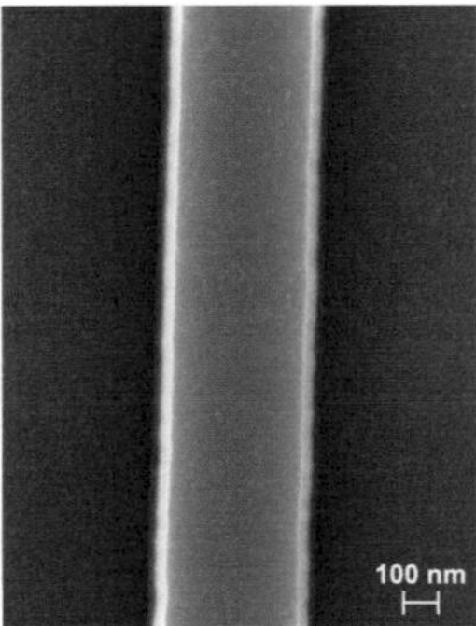

Fig. 4.2. SEM photograph of strip WG. The sidewalls exhibit roughness due to the limited resolution and accuracy of the lithography. White shadings can be observed at the WG boundaries, which indicate non-vertical sidewalls.

sidewalls. The transmission of a WG with a length of $1,250\,\mu$m is shown in Fig. 4.3, where WG tapers are included in the total length. Strong oscillations caused by reflections at the WG facets are clearly observed, and for clarity the moving average is added to the graph as red dashed line. The reflections could be removed by adding anti-reflection coatings at the facets, by using angled facets, or by coupling in light into the silicon WGs with inverted tapers or grating couplers. The transmission result represents the actually received power by the optical spectrum analyzer in dBm in a 2 nm bandwidth for a laser power of 0 dBm. The total transmis-

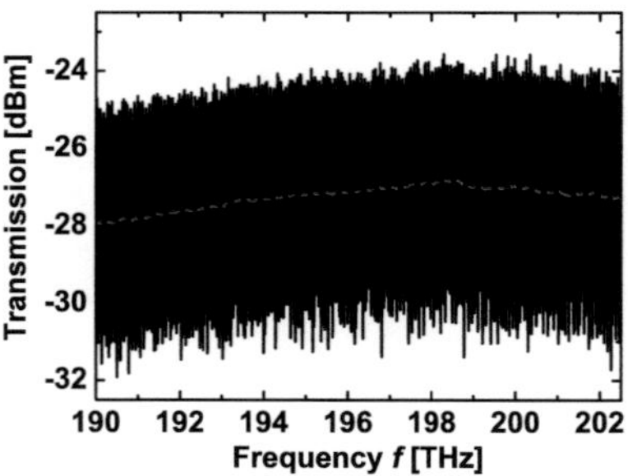

Fig. 4.3. Measured transmission of strip-WG. The red dashed line represents the moving average. The WG has a nominal width of 440 nm and a length of $1,250\mu$m including the WG tapers.

sion loss is the sum of the coupling loss from the fibers to the silicon WG, of the transmission loss of the WG, and of other additional losses in the setup like the loss of the polarizers, the fibers, the insertion loss of the switch and of the various connectors. By measuring a second strip WGs with a smaller device length of $1,000\,\mu$m and comparing the transmission, we can estimate the transmission losses of the WG, Fig. 4.4(a). We find a loss value of around 1 dB / mm for an actual WG width of around 415 nm, which compares well to other reported results using deep UV lithography [144]. In addition to the WG transmission losses, we also estimate the setup losses, which we define as the sum of the coupling losses and the additional setup losses, Fig. 4.4(b). The setup losses are estimated by subtracting the calculated transmission losses of the WG from the total transmission, and we find that the setup losses are approximately 26 dB. For the measurements shown in the next sections, the setup losses are expected to be similar. Thus, if the transmission characteristics of the PC-WGs with respect to strip-WG input and output ports need to be estimated, this might be done by subtracting the setup losses from the presented transmission results.

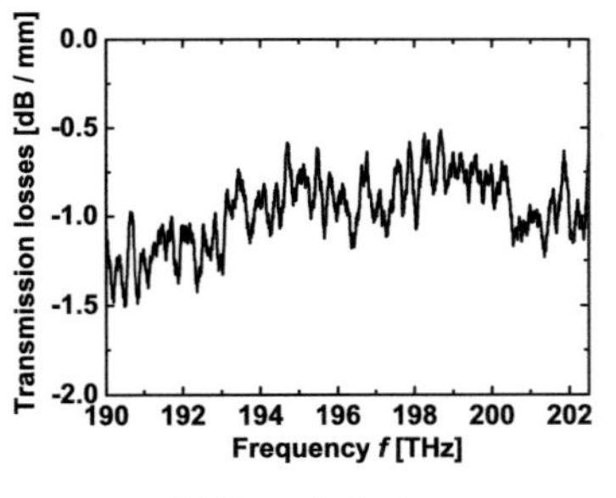

(a) Transmission losses.

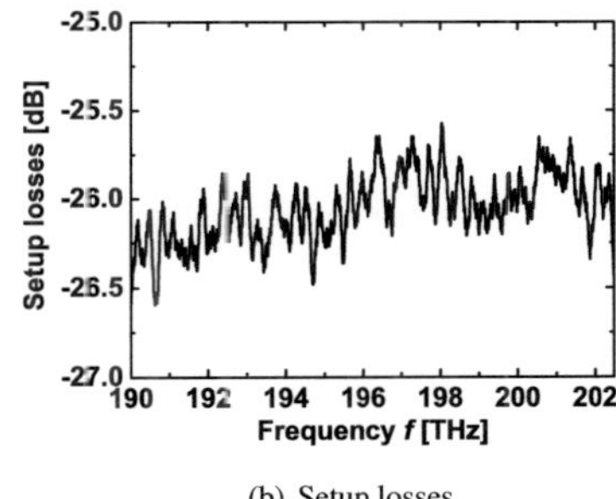

(b) Setup losses.

Fig. 4.4. Measured losses of strip WG. (a) The transmission losses are around 1 dB / mm and are determined by comparing the averaged transmission of two WGs of different lengths. (b) The setup losses are defined as the sum of the coupling losses from the fibers to the silicon WG and the additional losses of the setup. The setup losses are estimated to be around 26 dB by subtracting the transmission losses of the WG from its total transmission.

4.1.3 Broadband Slow Light Waveguide

The first PC device to verify experimentally is a broadband slow light PC-WG with a group velocity of 3.7 % of the vacuum speed of light in a THz bandwidth; the group velocity characteristics will be shown below in Fig. 4.7(b). In addition to the basic interest in such a specifically engineered WG, it is also part of the variable delay line proposed in Section 3.2. The design exactly corresponds to the structure 'Bur' of Subsection 2.2.3, which has a glass cladding. In the design procedure, the principles mentioned in Subsection 2.1.3 have been applied. The structure parameters of the PC-WG are $W_1 = 0.95\sqrt{3}a$, $W_2 = 0.9\frac{1}{2}\sqrt{3}a$, $W_3 = 0.95\frac{1}{2}\sqrt{3}a$, $r_1 = 0.25a$, $r_2 = 0.27a$, $r_3 = r = 0.3a$, and $a = 406$ nm. A SEM photo of the fabricated PC is displayed in Fig. 4.5. It can be clearly seen that the first three rows of holes next to the WG core have different hole radii, and we find that the structure parameters are very similar to the nominal design values. The holes are not perfectly round but exhibit roughness due to the limited resolution and accuracy of the lithography. White shadings at the hole boundaries indicate non-vertical sidewalls, and the small particles are remainders of process chemicals. At both ends of the slow-light PC-WG, PC tapers with a length of 7 periods are added to increase the

Fig. 4.5. SEM photograph of broadband slow-light WG. The first three rows of holes next to the WG core have different hole radii $r_1 = 0.25\,a$, $r_2 = 0.27\,a$, $r_3 = 0.3\,a$. The holes exhibit roughness due to the limited resolution and accuracy of the lithography. The small particles are remainders of process chemicals.

coupling to the strip-WGs, as explained in Subsection 2.1.4. Along the taper, the core width of the PC-WG is gradually increased to $W_1 = \sqrt{3}\,a$, while all other structure parameters are not varied. Three different lengths of PC-WG have been measured, which are $114\,a = 46.3\,\mu$m, $514\,a = 208.7\,\mu$m and $1014\,a = 411.8\,\mu$m, and sections of strip-WG and WG tapers are added. The measured transmission of the devices is displayed in Fig. 4.6. The transmission window is between 194 THz and 196.5 THz, and oscillations can be observed which are caused by reflections at the WG facets and at the boundary between strip-WG and PC-WG. The transmission losses can be estimated from the difference of the transmission of the three WG lengths, and the losses are minimum near 195.35 THz with a value of around 14.0 dB / mm.

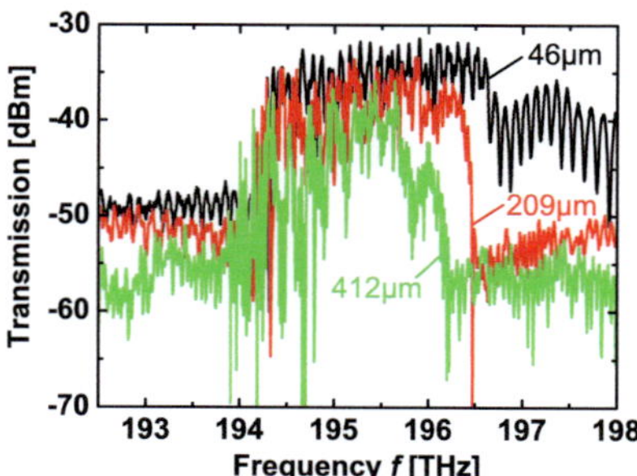

Fig. 4.6. Measured transmission of broadband slow light PC-WG. Three different PC-WG lengths with added sections of strip-WGs have been measured, and the minimal transmission losses are estimated to be around 14.0 dB / mm at 195.35 THz.

The measured group delays of the WGs of three different lengths are shown in Fig. 4.7(a). In addition to the measurement curves, the simulated group delays for the three

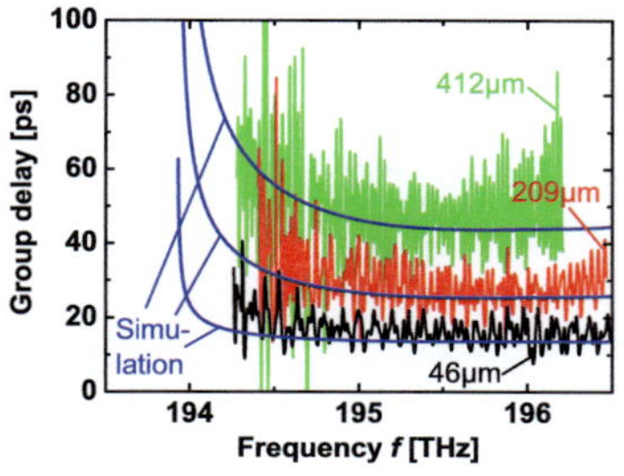

(a) Measured group delay compared to simulation.

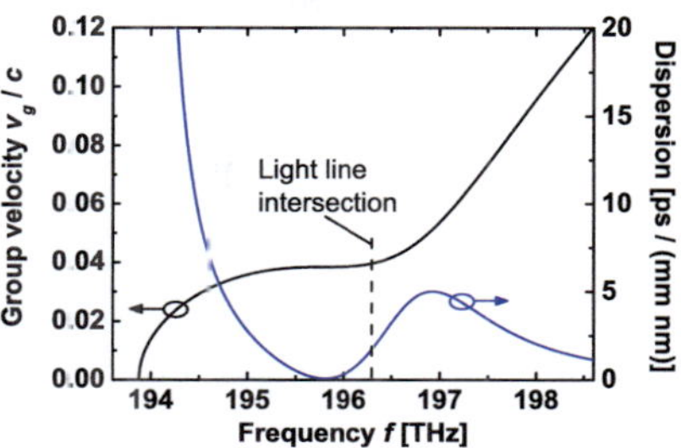

(b) Group velocity and dispersion.

Fig. 4.7. Measured group delay of broadband slow-light WG compared to simulation. (a) Three lengths of PC-WG are measured, and good agreement to simulations is achieved. (b) Group velocity and chromatic dispersion of PC-WG inferred from simulations. A group velocity of 3.7 % of the vacuum speed of light is found at zero chromatic dispersion.

WG devices are overlaid (blue thick lines), where the frequency shift between the measured and simulated curves has been adjusted such to match the data. It can be observed that the simulations fit the measurements, which verifies that the group delay characteristics of the fabricated PC-WGs behave as designed. The inferred group velocity and chromatic dispersion characteristics are displayed in Fig. 4.7(b). The broadband slow light behavior can be clearly seen, and near 195.75 THz the group velocity reaches 3.7 % of the vacuum speed of light while the chromatic dispersion almost vanishes.

4.1.4 Waveguide with Linearly Varying Chromatic Dispersion

The next experimentally verified slow-light PC-WG is designed to exhibit a linearly varying chromatic dispersion characteristics and a dispersion minimum that is negative; the group velocity characteristics will be shown below in Fig. 4.10(b). Such a PC-WG is the basic part of the tunable dispersion compensator and of the tunable delay line proposed in the Sections 3.1 and 3.2. The design with added glass cladding has already been presented in Subsection 3.1.2, but for the fabricated structure the period a is slightly changed. The structure parameters are $W_1 = 0.95\sqrt{3}\,a$, $W_2 = \frac{1}{2}\sqrt{3}\,a$, $W_3 = 0.9\frac{1}{2}\sqrt{3}\,a$, $r_1 = 0.25\,a$, $r_2 = 0.32\,a$, $r_3 = 0.34\,a$, $r = 0.3\,a$, and $a = 417$ nm. A SEM photo of a PC section is displayed in Fig. 4.8. As for the broadband slow-light WG, the structure parameters are again very similar to the nominal design values, and the holes exhibit fabrication-induced roughness. At both ends of the PC-WG, PC coupling tapers with a length of 7 periods are added, along which the core width is again increased to $W_1 = \sqrt{3}\,a$. The three different lengths of PC-WG that have been measured are $114\,a = 47.5\,\mu$m, $514\,a = 214.3\,\mu$m and $1014\,a = 422.8\,\mu$m, and additional strip-WG sections are included in the results. The measured transmission characteristics of the devices are displayed in Fig. 4.9. The transmission window is between 191 THz and 194.5 THz, and again oscillations can be observed which are caused by reflections. The estimated transmission losses are minimal near 191.85 THz with a value of around 11.15 dB / mm. At certain spectral posi-

Fig. 4.8. SEM photograph of WG with linearly varying chromatic dispersion. The first three rows of holes next to the WG core have different hole radii $r_1 = 0.25\,a$, $r_2 = 0.32\,a$, $r_3 = 0.34\,a$. The holes exhibit roughness due to the limited resolution and accuracy of the lithography. The small particles are remainders of process chemicals.

tions, the transmission drops considerably, e. g. at 192.67 THz, 193.02 THz and 193.36 THz. We attribute this to coupling of the TE mode to lossy TM modes. The coupling is caused by non-vertical sidewalls of the fabricated structure, which introduces a vertical structure asymmetry. The issue of TE-TM mode coupling has been discussed in more detail in Subsections 1.4 and 2.2.4, and has been found by experiments [147], [27].

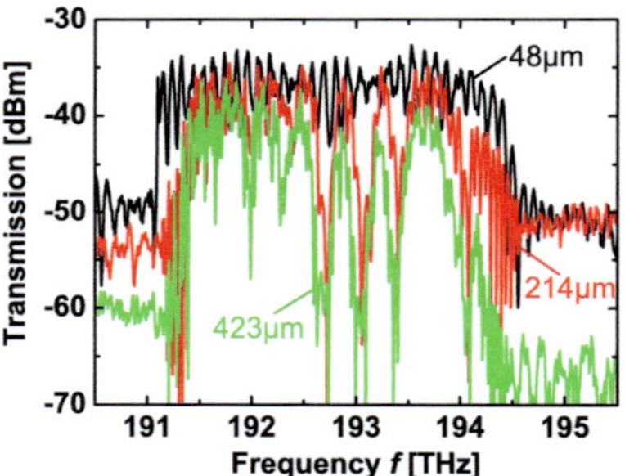

Fig. 4.9. Measured transmission of PC-WG with linearly varying dispersion. Three different PC-WG lengths with added sections of strip-WGs have been measured, and the minimal transmission losses of the PC-WG are estimated to be around 11.15 dB / mm at 191.85 THz.

The measured group delays of the WGs with three different lengths are shown in Fig. 4.10(a). Again, good agreement is found between the simulated group delay for the three WG lengths (blue thick lines) and the measurements for a suited frequency shift between the measured and simulated curves. The corresponding simulated group velocity and chromatic dispersion characteristics of the devices are shown in Fig. 4.10(b). The chromatic dispersion is linearly decreasing with frequency up to a frequency of 194.1 THz, where the dispersion is minimal with a negative value of $-5\,\mathrm{ps}/(\mathrm{mm\,nm})$; for higher frequencies the dispersion increases again. The group velocity at the frequency of minimum dispersion is 3.6 % of the vacuum speed of light.

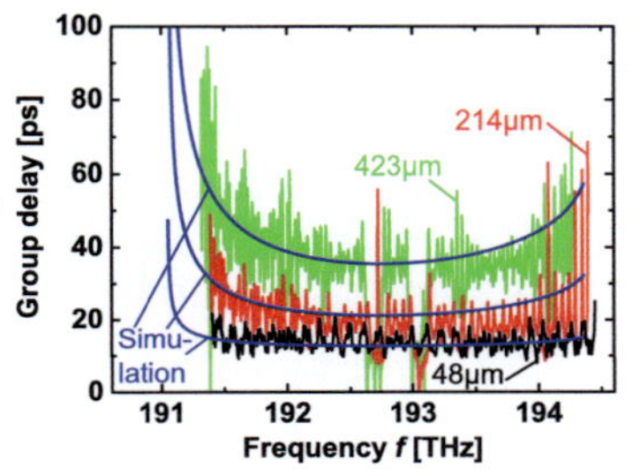

(a) Measured group delay compared to simulation.

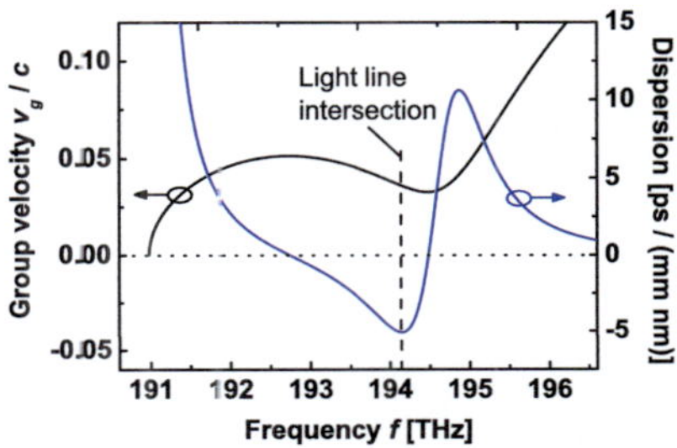

(b) Group velocity and dispersion.

Fig. 4.10. Measured group delay of PC-WG with linearly varying dispersion compared to simulation. (a) Three lengths of PC-WG are measured, and good agreement to simulations is achieved. (b) Group velocity and chromatic dispersion of PC-WG inferred from simulations. The minimal chromatic dispersion reaches $-5\,\mathrm{ps}/(\mathrm{mm\,nm})$, and the group velocity at the corresponding frequency is 3.6 % of the vacuum speed of light.

4.1.5 Slot Waveguide

Finally, we show the measurement results from a fabricated slow-light slot PC-WG. Such a PC-WG is the basis of the fast electro-optical modulator proposed in Section 3.3, and a high electro-optic interaction can take place inside the narrow slot if a nonlinear material is infiltrated or grown. To verify the basic linear characteristics of the WG, a glass cladding has been grown on top of the silicon structure. In the first realization presented here, the group velocity or dispersion of the slot PC-WG has not been specifically engineered to have constant group velocity or linear dispersion, but starting at the band-edge the group velocity increases monotonously with frequency and the chromatic dispersion decreases with frequency, Fig. 4.13(b). The principle of the WG and more details on how to obtain specific dispersive properties have been presented in Sections 2.1 and 3.3. The nominal structure parameters of the slot PC-WG are $W_{\mathrm{gap}} = 0.135\,\mu\mathrm{m}$, $W_1 = 1.554\sqrt{3}\,a$, $W_2 = W_3 = \frac{1}{2}\sqrt{3}\,a$, $r_1 = r_2 = r_3 = r = 0.3\,a$, and $a = 424\,\mathrm{nm}$. A SEM photo of the fabricated slot PC-WG is show in Fig. 4.11(a). For the chosen fabrication process parameters, the realized gap width $W_{\mathrm{gap}} \approx 0.18\,\mu\mathrm{m}$ came out larger than designed and the hole radii $r \approx 0.285\,a$ smaller than designed. If the lithography mask is appropriately predistorted, it can be achieved that both the hole size and the gap width are fabricated correctly. In order to couple light in into the slot PC-WG, first the feeding strip-WG is transformed into a slot-WG, which then is coupled to the slow-light PC-WG. The coupling structure is described in more detail in Subsection 3.3.6, and a SEM photograph is shown in Fig. 4.11(b). It can be seen that the gap in the strip-WG to slot-WG transition vanishes completely; this could be improved in a next fabrication run by appropriately changing the lithography mask. In the modulator design, Fig 3.16, the two strips have been separated to electrically isolate the silicon parts that will later be doped. However, even if the two strips are connected at the transition, we estimate that

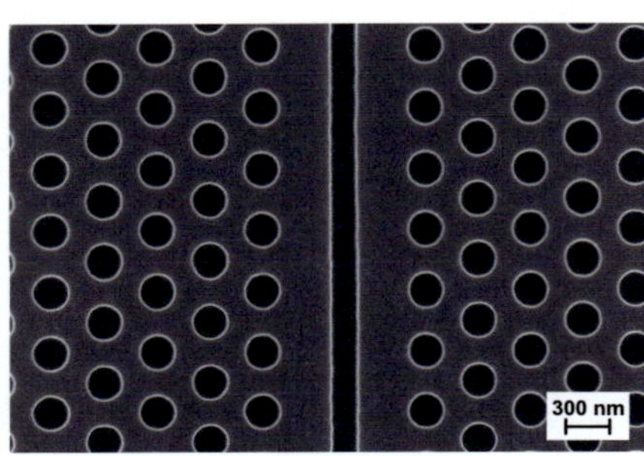

(a) Section of slot PC-WG.   (b) Strip-WG to slot WG transition.

Fig. 4.11. SEM photograph of slow-light slot WG and of coupling structure. (a) The holes are fabricated smaller than designed, and the slot larger. In addition, the holes and the slot exhibit roughness due to the limited resolution and accuracy of the lithography. (b) In the strip-WG to slot-WG transition, the gaps are not fabricated correctly, which degrades the device performance.

the relatively low doping level and the small strip width lead to a very high resistance of the connection. At each end of the slow-light PC-WG, PC tapers are again added to increase the coupling to the slow-light mode; the taper has a length of 7 periods and the core width is increased to $W_1 = 1.6\sqrt{3}\,a$. Three PC-WG with different lengths have been measured, which are

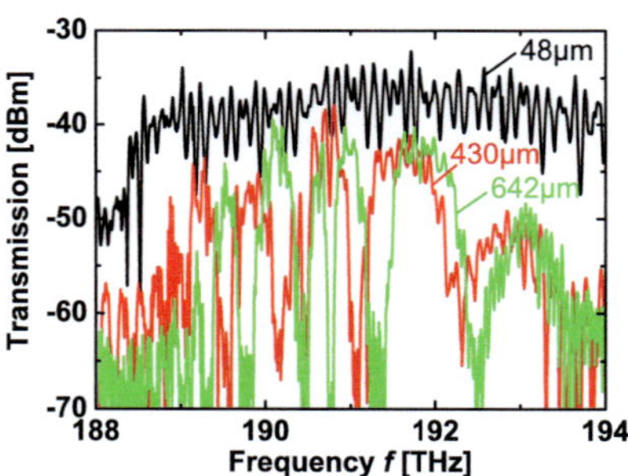

Fig. 4.12. Measured transmission of slot PC-WG. Three different PC-WG lengths with added sections of strip-WGs have been measured, and the minimal transmission losses of the PC-WG are estimated to be around 10 dB / mm at 191.5 THz.

$114\,a = 48.3\,\mu\text{m}$, $1014\,a = 430\,\mu\text{m}$ and $1514\,a = 642\,\mu\text{m}$, and connecting strip-WG sections are added. The measured transmission of the devices is displayed in Fig. 4.12. The transmission window is between 189 THz and 193 THz, and the oscillations are due to reflections at the WG facets and at the boundary between strip-WG and PC-WG. The estimated transmission losses are minimal at 191.5 THz with a value of around 10 dB / mm. At certain spectral positions, the transmission again drops considerably. As for the WG with linearly varying dispersion, this can be attributed to coupling of TE and TM modes caused by non-vertical sidewalls.

The measured group delays of the WGs with three different lengths are shown in Fig. 4.13(a). Again, good agreement is found between the simulated group delay for the three WG lengths and the measurements for a suited frequency shift between the measured and simulated curves. In the simulations, the structure parameters have been changed to the fabricated values. The corresponding group velocity and chromatic dispersion characteristics of the devices are shown in Fig. 4.13(b). The group velocity varies between 0 % and 12.3 % of the vacuum speed of light below the light line, and the chromatic dispersion decreases with increasing group velocity.

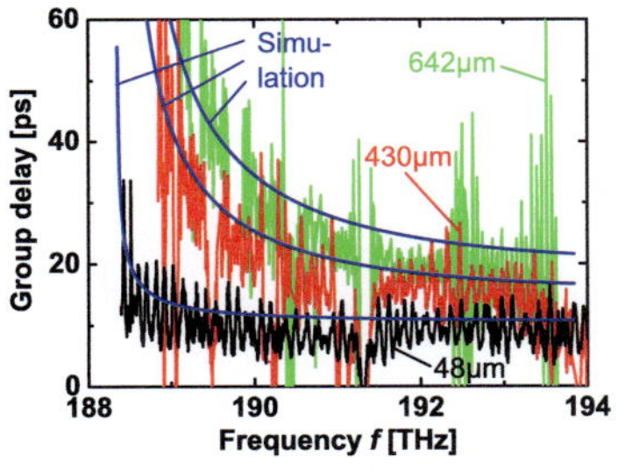

(a) Measured group delay compared to simulation.

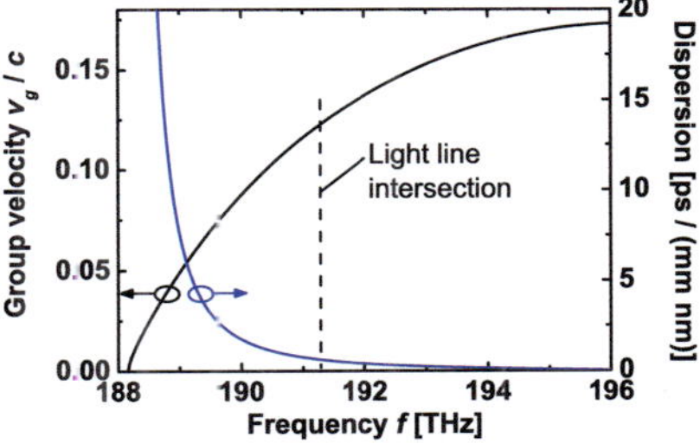

(b) Group velocity and dispersion.

Fig. 4.13. Measured group delay of slot PC-WG. (a) Three lengths of PC-WG are measured, and good agreement to simulations is achieved. (b) Group velocity and chromatic dispersion of PC-WG inferred from simulations. The group velocity varies between 0 % and 12.3 % of the vacuum speed of light below the light line, and the chromatic dispersion decreases with increasing group velocity.

4.1.6 Conclusion

We design and characterize slow-light PC-WGs fabricated with deep-UV lithography. A broadband slow-light WG with a group velocity of 3.7 % of the vacuum speed of light shows minimum transmission losses of 14.0 dB / mm. A WG with linearly varying chromatic dispersion and a dispersion minimum of $-5\,\text{ps}/(\text{mm}\,\text{nm})$ has minimum losses of 11.15 dB / mm, and a slow-light slot-WG exhibits minimum losses of 10 dB / mm. For all three PC-WGs, the measured group delay agrees nicely with the simulated delay, which verifies the design. However, transmission dips occur in the spectra, which can be attributed to mode-coupling caused by non-vertical sidewalls. By improving the fabrication process, the mode coupling and the transmission losses can be decreased.

4.2 Microwave Experiments

Next generation nano-photonic components are both difficult to fabricate and to simulate. Device scaling into the microwave regime allows to experimentally verify novel devices and to proof the design concepts. So far, free-space microwave experiments were conducted, which provided the properties of bulk photonic crystals (PCs) [5], [148], [149], [150] and of 2D PC devices [151], [152], showed photon localization in disordered systems [153], and revealed a negative index of refraction in metamaterials [154] as well as negative refraction in PCs [155], [156], [157]. Yet free-space microwave experiments are not well suited to validate guided-wave photonic crystals with high index contrast, as vacuum wavelengths much larger than typical structure dimensions lead to source coupling problems and to spurious radiation. However, guided-wave photonic crystals [9] are of special interest because they offer unique possibilities to control dispersion [14], [158] and to slow down the flow of light [14], [88], while they can have a high bandwidth as needed for application in fast optical communication systems. Slow-light PC structures exhibit an enhanced light-matter interaction [15] which can significantly reduce the size and control power needed for optical modulators [60], [16] and nonlinear elements [113], and they might also be candidates for optical buffers and delay lines [67]. Various broadband slow-light devices have been suggested [26], [87], [159] and realized [93], [95]. Devices with high dispersion could be used for residual-dispersion compensation [39]. For a potential application of the structures, the influence of fabrication-induced disorder on the device performance must be known, and several numerical approaches have been proposed for investigation [22], [52], [57], [59].

While there is an obvious need for accurate characterization of such devices, the exact properties are difficult to access in the optical regime, and the accuracy of the numerical design tools and of the disorder models are hard to classify. This is because in optical experiments, the influence of fabrication-induced disorder is always present and cannot be isolated, and the accuracy of optical measurement equipment is often a limiting factor.

In this Section, the properties of guided-wave photonic devices with high index contrast are experimentally proven in the microwave regime. This enables us to study the device characteristics in detail and to verify the results of numerical tools. As an example, a broadband slow-light or high-dispersion PC-WG is studied. In the microwave model, structures can be fabricated with an equivalent optical accuracy of 0.5 nm, which allows us to investigate disorder which is introduced in a controlled manner. Measurement equipment at microwave frequencies is very accurate and covers octave-band frequency ranges, so that one can characterize components in great detail. The microwave model configuration is flexible and modular; therefore we can easily remove or concatenate different WG sections.

The section is organized as follows: In Subsection 4.2.1, the microwave model is introduced, and the microwave materials and the mode excitation scheme are discussed. In Subsection 4.2.2, the broadband PC-WG is presented, and the results of the measurements and simulations are discussed. Also, disordered structures are treated. Details about the different microwave measurement techniques are presented in Appendix A.8, and the mathematics of the used TRL calibration are given in Appendix A.4.

Parts of this section have been published in a journal article, [2] or [J2] on page 165.

4.2.1 Microwave Model

For the microwave experiments, we enlarge the structure by a factor of 20,000 and diminish the frequency by the same amount from around 200 THz to 10 GHz, which is possible because of the scaling properties of Maxwell's equations. At these microwave frequencies, equivalent dielectric materials with high refractive index are available. To efficiently excite the mode in a dielectric strip WG, we have designed a broad-band slot antenna. Highly accurate measurement equipment is available, and we use three different measurement techniques to characterize the devices: Continuous wave measurements, pulse measurements, and near-field measurements, with details given in Appendix A.8.

Dielectric Microwave Materials

Our aim is to study devices with a high-index contrast, similar to silicon on insulator (SOI) in the optical regime. Table 4.1 lists the properties of some suitable microwave materials. The loss factor is defined as $\tan\delta = \varepsilon_r/\varepsilon_{ri}$, where the complex permittivity has the real part $\varepsilon_r = n^2$ and the imaginary part $-\varepsilon_{ri}$.

	C-Stock AK 10	TMM 10	Al_2O_3	RT 5880	PE-HD	C-Stock RH-5
ε_r	10 ± 0.5	9.2 ± 0.23	9.9	2.2 ± 0.02	2.3	1.09
n	3.16	3.03	3.15	1.48	1.52	1.04
$\tan\delta$	0.002	0.0022	0.0002	0.0009		0.0004

Table 4.1. Properties fo different microwave materials at 10 GHz. At optical frequencies near 193 THz, the refractive index of silicon $n = 3.48$ and of glass $n = 1.44$. The loss factor of glass is $\tan\delta = 6 \times 10^{-12}$.

For the high-index material with a refractive index above 3, C-Stock AK 10, TMM 10 or alumina (Al_2O_3) can be used. C-Stock AK from Cuming Microwave Corporation and TMM from Rogers Corporation are both ceramic-filled plastics, which are available with different permittivities. We have chosen C-Stock AK 10 for our experiments, as it can be easily handled and machined with standard milling tools, and as the material shows good isotropy, which will be presented below. TMM 10 is more brittle and thus more difficult to be milled, whereas alumina is so hard that it can only be machined with diamond tools, although this choice would be most desirable with respect to material losses.

The substrate material needs to have a refractive index around 1.5, where RT 5880 from Rogers Corporation, and PE-HD (high-density polyethylene) are two possible choices. To model a membrane structure, a foam material with a refractive index near 1 is used as substrate, like C-Stock RH-5 from Cuming Microwave Corporation.

The losses of microwave materials are higher than for equivalent optical materials, which has to be taken into account. Also, material nonlinearities are negligible, which means that nonlinear effects can not be studied.

Excitation of the Waveguide Mode

The devices under investigation are waveguide-based, and we use a dielectric strip-WG with only one mode for each polarization near 10 GHz as the basic component of the experiments. The WG is excited with a broadband slot antenna, which has a high polarization purity.

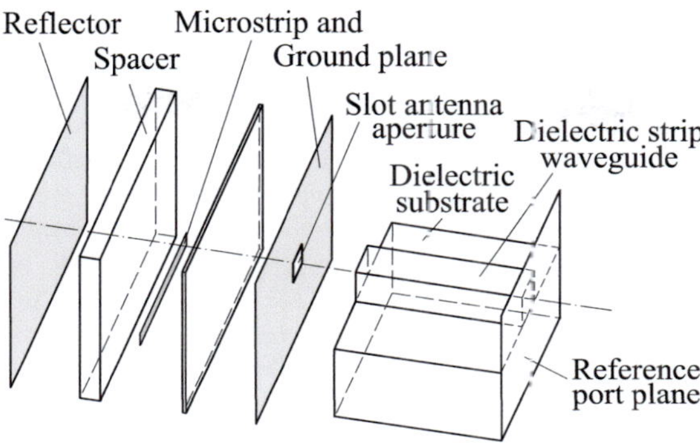

Fig. 4.14. Exploded view of the microstrip-fed slot antenna. Metallic sheets are shaded in gray. Tight coupling between slot antenna and dielectric strip-WG provides low reflection and prevents spurious radiation. The dielectric strip-WG has a width of 5.4 mm and a height of 2.7 mm.

The antenna is shown in an exploded view in Fig. 4.14. It consists of a metallic reflector plate preventing backward radiation, a spacer made of dielectric foam, and a standard RT 6010 microstrip substrate carrying a 50 Ω microstrip line that excites the slot in the ground plane. A photograph of the antenna where the dielectric strip waveguide has been detached is displayed in Fig. 4.15. At the end of the microstrip line, an SMA connector is attached, which connects the antenna to the measurement equipment via a coaxial cable.

Fig. 4.15. Broadband slot antenna to excite the fundamental dielectric strip-WG mode near 10 GHz. For clarity, the strip-WG is detached from the antenna.

Material Characterization

To characterize the high-index microwave material and especially verify its permittivity, we design and fabricate a single-moded strip-WG of the material. The dispersion relation of this WG can be inferred from microwave measurements, and the fit to simulations allows estimating the deviation of the material permittivity from its nominal value. There are two possibilities to measure the effective refractive index n_{eff} of the WG for a certain frequency.

Firstly, one of the dominant field components can be measured with the near-field measurement setup along the WG direction, shown in Fig. A.7 of Appendix A.8. To obtain a standing wave inside the WG, as needed to observe intensity variations, a reflecting copper plate is positioned at the WG end. The measurement curve exhibits maxima and minima, and the period corresponds to half the modal wavelength. The effective refractive index n_{eff} of the mode can be retrieved from the curve either by evaluating the peaks of the spatial Fourier transform or by fitting it to the theoretically expected trace $[A_1 \cos(2\pi n_{\mathrm{eff}} f/c + \varphi_1)]^2$, where f is the operating frequency, c the vacuum speed of light, and A_1 and φ_1 fitting parameters for the amplitude and phase. For higher operating frequencies, also a higher order mode might be excited, which leads to a beating in the resulting curve. The effective refractive indexes of both modes can be evaluated by fitting to $[A_1 \cos(2\pi n_{\mathrm{eff},1} f/c + \varphi_1) + A_1 \cos(2\pi n_{\mathrm{eff},2} f/c + \varphi_2)]^2$.

Secondly, the effective refractive index can also be retrieved from the phase information of the CW transmission experiment, see Appendix A.8. From the phase $\varphi = -\beta L = -2\pi n_{\mathrm{eff}} f/c\, L$, the effective refractive index can be directly calculated as a function of frequency. However, a reference measurement with the near-field setup is needed at one frequency, as the phase φ from the CW measurements has an unknown offset.

For the material TMM10 having a nominal refractive index of $\varepsilon_{\mathrm{r}} = 9.2$, we design a square WG with width and height $w = h = 7.3\,\mathrm{mm}$, which is fabricated and measured with the near-field setup. As WG substrate, a foam material with similar characteristics to air is used. The extracted dispersion diagram is compared to the simulation in Fig. 4.16(a). The measured curve

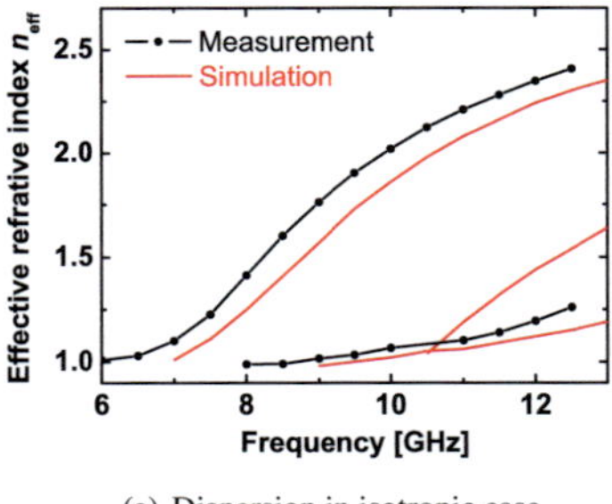

(a) Dispersion in isotropic case.

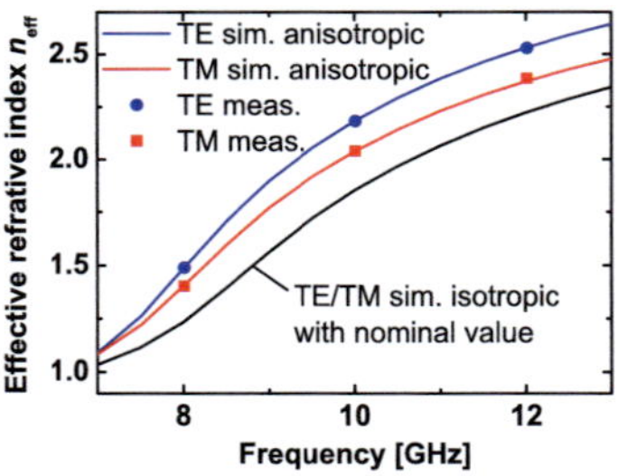

(b) Dispersion in anisotropic case.

Fig. 4.16. Measured effective refractive index of a strip-WG compared to simulation results. (a) A shift between the TM measurement and the isotropic simulation with nominal permittivity is observed caused by a deviation of the actual material permittivity. (b) A difference in effective refractive index for TE and TM polarization occurs. By a simulation with an anisotropic material, the behavior can be well reproduced using $\varepsilon_{\mathrm{r,x}} = 10.77$ and $\varepsilon_{\mathrm{r,y}} = 9.6$.

is shifted in frequency of about 0.5 GHz with respect to the simulated curve, which indicates a deviation in the material permittivity of TMM10 from its nominal value. As the WG is of square shape, the TE and TM modes should theoretically have exactly the same dispersion relation. However, we find that a significant difference occurs between the TE and TM mode dispersion, as shown in Fig. 4.16(b). This difference can only be explained by a material anisotropy, which means that there are different permittivity values for the two orthogonal transverse electric field directions x and y. By modeling a WG with an anisotropic material also in the simulations, the measurement results can be better reproduced, and with $\varepsilon_{r,x} = 10.77$ and $\varepsilon_{r,y} = 9.6$ we find a good match.

We also fabricate and measure strip-WGs with the material CAK-10, and the measurements show a negligible anisotropy, but also significant deviations from the nominal permittivity. For different material lots, we infer relative permittivities ε_r between 8.9 and 10.1. The isotropy of CAK-10 is another advantage of the material in addition to the good machinability, and the material is used for the following experiments. A single-moded WG is designed with a width $w = 8.1$ mm and a height $h = 5.4$ mm.

4.2.2 Broadband Slow Light Waveguide

Numerous optical components can be studied with the microwave model, like waveguide couplers, ring resonators and photonic crystals. We characterize a slow-light or high-dispersion photonic crystal waveguide in detail, which exhibits a group velocity of 4 % of the vacuum speed of light over a bandwidth of more than 1 THz, and a region with a chromatic dispersion of 4 ps / (mm nm) over 1 THz. The results of the CW, pulse transmission and near field measurements are presented, and used to verify the accuracy of the FIT simulation method. Also, the delay-bandwidth product of the WG is estimated. For a real device, fabrication-induced disorder is unavoidable, and we show experimental results of the influence of 5 % radial disorder.

Structure and Design

Following a recent proposal [26] we designed the single-moded broadband slow-light PC-WG as an air-bridge structure for a slab permittivity of $\varepsilon_r = 10$. A triangular lattice PC was chosen having a slab thickness $h/a = 0.6$, hole radii $r/a = 0.25$, and a reduced width W0.75 PC-WG. The period of the PC is $a = 0.45 \, \mu$m in the optical regime and $a = 9$ mm for the microwave model. Fig. 4.17 shows the band diagram of the structure, generated with the FIT eigenmode solver. The region with low group velocity is indicated in the figure as "Slow light region"; it is centered around a frequency of 205 THz. To efficiently couple the PC-WG to a feeding strip-WG, a PC taper is introduced, where the width of the PC-WG is adiabatically increased from W0.75 to W0.8 over a length of 5 periods [26]. The structure is schematically shown in Fig. 4.18(a). The field distribution of the slow-light mode has considerable field values in the region of the air holes. Along the taper this field is transformed to be confined mainly to the waveguide core for a better match to the field distribution of the strip WG. Simulations with the FIT transient solver indicate a coupling loss below 1.25 dB per interface and a reflection below 10 dB in the slow-light regime, which are much better values than for direct coupling to the strip-WG without taper [100]. We fabricated a microwave model as shown in Fig. 4.18 with coupling tapers and 30 periods of the W0.75 PC-WG.

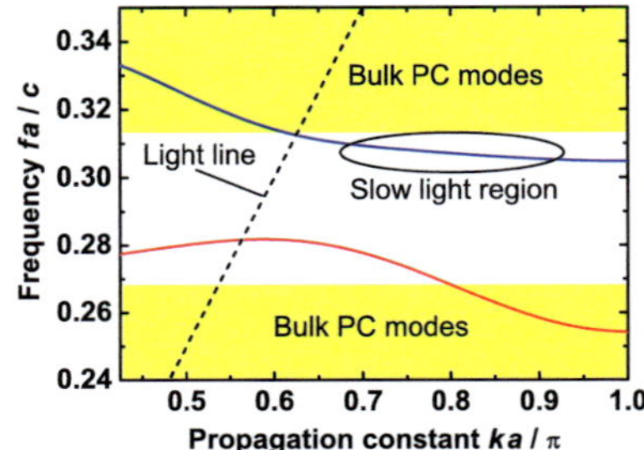

Fig. 4.17. Band diagram of an air bridge W0.75 PC-WG designed to have a slow light region with a flat band shape. The group velocity is 4% of the vacuum speed of light in an 1.2 THz bandwidth at a central frequency of 204.7 THz.

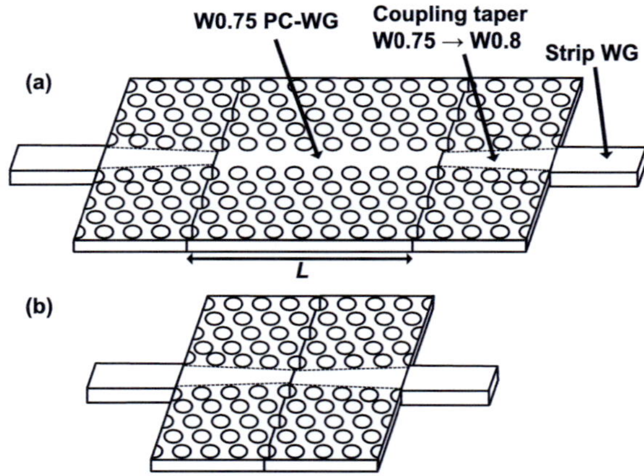

Fig. 4.18. (a) Schematic of the PC device comprised of three different building blocks: A feeding strip-WG, a PC coupling taper, and a W0.75 PC-WG of length *L*. The different blocks can be separated or plugged together. The tapers improve the coupling between strip-WG mode and slow-light PC mode. (b) For reference measurements, the W0.75 PC-WG block can be removed and the coupling tapers can be directly connected.

Device Characteristics

The properties of the device are characterized with CW measurements, more details on the experimental technique are found in Appendix A.8. Numerical simulations with the FIT transient solver have been conducted, where also material losses were considered. Transmission and reflection of the complete PC device as shown in Fig. 4.18(a) are displayed in Fig. 4.19. Near

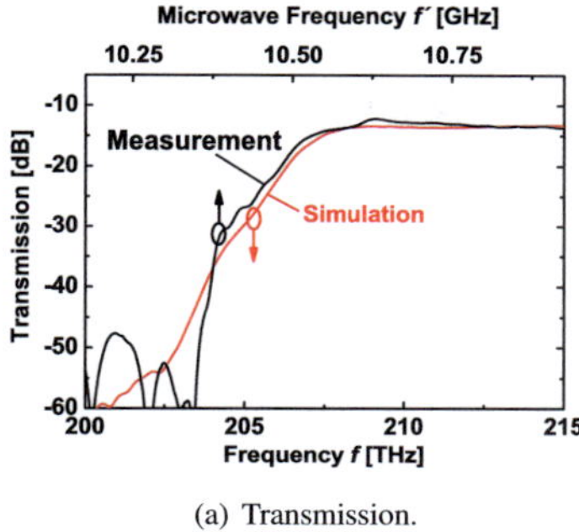

(a) Transmission.

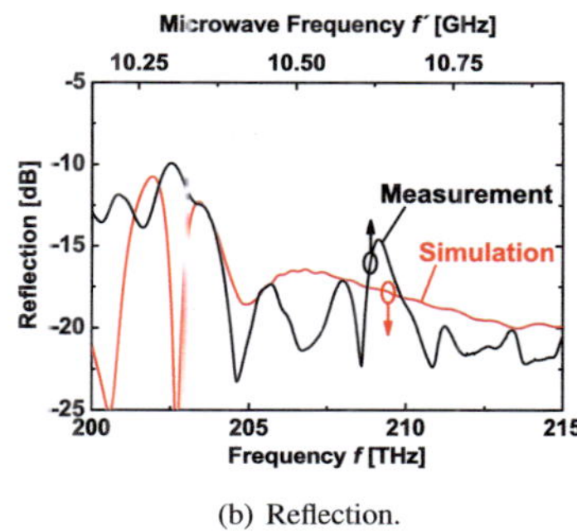

(b) Reflection.

Fig. 4.19. Transmission and reflection spectra of the PC device according to Fig. 4.18(a). Results from microwave measurements and from FIT simulations agree well. Arrows point to the respective frequency axes.

$f = 203\,\text{THz}$, the band cutoff is reached, and for lower frequencies the transmission drops considerably while the reflection is increased. The measured and simulated curves agree well. In the reflection graph, the traces do not exactly match, but rather the curve envelopes. We find that the exact oscillations are rather hard to predict, as the reflected power level is low and small phase changes can shift the position of the maxima and minima strongly.

Fig. 4.20 display the group velocity and chromatic dispersion for the de-embedded PC WG (inner section in Fig. 4.18(a)), where the taper influences have been eliminated. The broadband region where the group velocity is around 4 % of the speed of light is centered near $f = 204.7\,\text{THz}$, and here the chromatic dispersion has a minimum. The bandwidth where the group velocity varies about $0.034 \times c$ with a deviation of $\pm 10\%$ is 1.2 THz, which is a remarkably large value. The measured group velocity curve in Fig. 4.20(a) shows ripples for low group

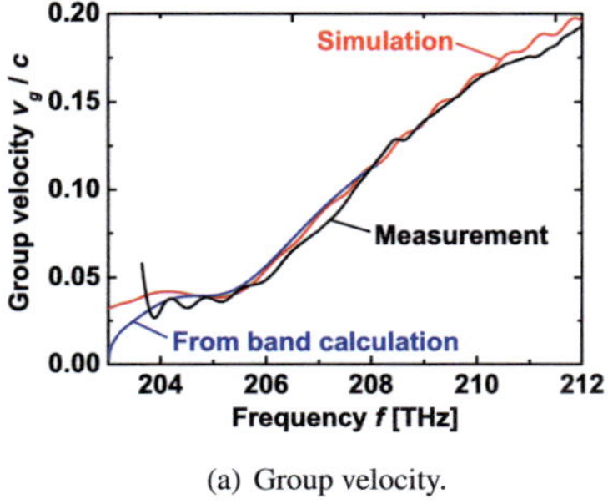

(a) Group velocity.

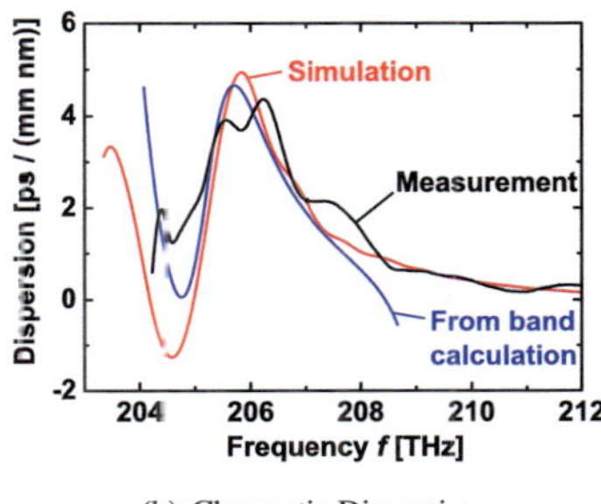

(b) Chromatic Dispersion.

Fig. 4.20. Group velocity and chromatic dispersion behavior of the PC-WG (inner section in Fig. 4.18(a)). Scaled microwave measurements, FIT simulations, and band calculation results show good agreement. Near $f = 204.7\,\text{THz}$, the group velocity reaches 4 % of the vacuum speed of light in an excellent 1.2 THz bandwidth. Near $f = 206\,\text{THz}$, a high chromatic dispersion value of 3.9 ps / (mm nm) in a broad 1.1 THz range is reached.

velocity values, which may be explained by Fabry-Perot resonances generated by reflections in the microwave model at the interfaces between coaxial cable and strip waveguide. In a 1.1 THz range around $f = 206$ THz, the chromatic dispersion is 3.9 ps / (mm nm) with a tolerance of $\pm 10\%$. This means that 1 mm of the PC-WG generates the same dispersion as 230 m of a standard single mode fiber, which has a dispersion of 17 ps / (mm nm).

In addition to the scaled microwave CW measurement results and the FIT transient simulation data, the curves obtained from FIT band diagram calculations are also shown in Fig. 4.20. Again, the FIT simulation results agree well with the measured data. It has to be noted that the FIT band calculations become unreliable if the light line is exceeded, which is near 208.2 THz, and so the curve ends there.

Pulse Transmission and Delay-Bandwidth Product

To study the feasibility of the component in a broadband communication system, we also conducted pulse transmission experiments in the microwave regime, more details on the experimental technique are found in Appendix A.8. The full width at half maximum (FWHM) equivalent optical bandwidth of the pulse is 1.3 THz. For the pulse measurements we use a cut-back method of the following kind: We first measure the transmission through a long WG ("output pulse"), and then cut out a WG section and refer to the transmitted signal as "input pulse". The difference in propagation times corresponds to the propagation time through the removed section. This procedure was applied to the PC-WG, where we cut out the section of 30 periods of W0.75-WG with length L, Fig. 4.18, and to a strip-WG; where we cut out the same length L. The four measured pulses are displayed in Fig. 4.21(a), where the two input pulses were adjusted to coincide by choosing appropriate reference times. The time axis is scaled to the optical regime, and the pulses are normalized with respect to the energy. The output pulse of

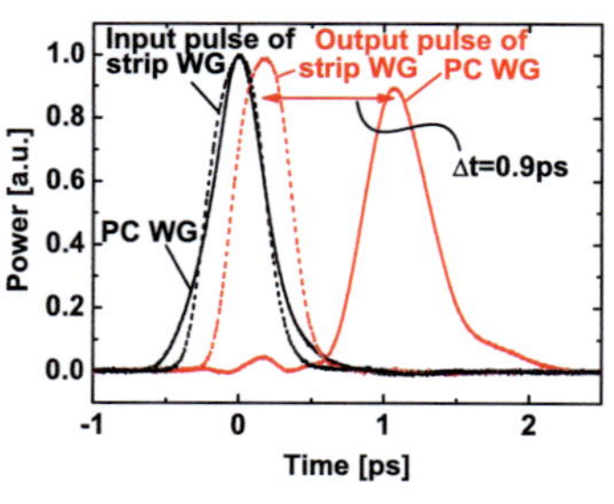

(a) Delay of a broadband pulse.

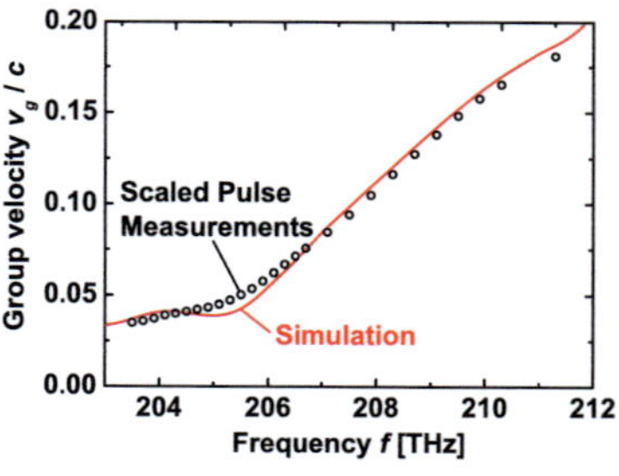

(b) Group velocity.

Fig. 4.21. (a) Delay of a broadband pulse with an equivalent optical pulse bandwidth of 1.3 THz in a PC-WG. Cut-back transmission measurements through a PC-WG and a strip-WG, where the output pulses have propagated through long WGs, and the input pulses through WGs with a cut-out section of length L. The time axis is scaled to the optical regime, and the reference times are adjusted such that the two input pulses coincide. (b) Group velocity for the PC-WG of Fig. 4.18(a), obtained from the measured time difference of the centers of broadband pulses. Good agreement with simulations is observed.

the PC-WG arrives significantly later than the output pulse of the strip-WG. It can also be observed that the output PC-WG pulse is broader than the input PC-WG pulse, which is due to the residual chromatic dispersion of the PC-WG in the slow-light regime. In Fig. 4.21(b), the group velocity as a function of pulse carrier frequency is displayed. The group velocity is obtained from the time difference of the input and output pulse centers. Assuming a measured pulse shape $P(t)$, the pulse centers are determined by $\langle t \rangle = \int_{-\infty}^{+\infty} t\, P^2(t)\, dt \Big/ \int_{-\infty}^{+\infty} P^2(t)\, dt$. The measurement results are again compared to FIT transient solver simulations, and good agreement is found.

We estimate the delay-bandwidth product of the PC-WG, which is a measure of how many optical information packets can be maximally stored on it. It is defined as the product of storage time T_S and packet rate B_{packet} [67]:

$$C_{\text{dly-bw}} = T_S\, B_{\text{packet}}. \tag{4.1}$$

The storage time for a PC-WG of length L is $T_S = L/v_g$, and the packet rate for on-off-keying with a bit rate f_T is $B_{\text{packet}} = f_T$. We set the maximal length of the PC-WG to a value so that the pulse broadening by chromatic dispersion is less than one third of the bit duration, where we use the linear law for pulse broadening $\Delta t_g = C_\lambda\, L \Delta \lambda$ with chromatic dispersion C_λ and spectral bandwidth $\Delta \lambda$. From simulations without material losses, we calculate a delay-bandwidth product $C_{\text{dly-bw}} \approx 1$ for a bandwidth of $1.3\,\text{THz}$ and $C_{\text{dly-bw}} \approx 150$ for a bandwidth of $125\,\text{GHz}$, e.g. for a $40\,\text{Gbit/s}$ optical signal. For the latter bandwidth, however, the PC-WG needs to be very long ($\approx 45\,\text{mm}$), and the delay-bandwidth product will in that case possibly be limited by propagation losses.

Field Distribution

For propagation at low group velocity, the field distribution inside the PC-WG is considerably different than for propagation at higher group velocities. We measured several components of the field amplitudes with the setup shown in Appendix A.8 at two frequencies and compared them to the field distribution obtained from FIT eigenmode calculations, Fig. 4.22. The measurements confirm the simulations very well. For propagation at low group velocity, the fields extend considerably into the patterned PC region, Fig. 4.22(a). This is the field distribution of the so-called gap-guided mode. The power flows not only in the center of the WG, but also in the region of the holes. Between the first and second row of holes next to the WG, the power even flows in opposite direction of the wave propagation. For frequencies where the mode exhibits a higher group velocity, the field is more confined to the PC-WG core, Fig. 4.22(b). This is the reason, as already explained in Subsection 4.2.2 and Subsection 2.1.4, why it is much more efficient to couple with a strip WG to the mode at high group velocity. The power flows only in the WG center.

Disorder Influence

In all real optical PC devices, the structures cannot be perfectly fabricated, but always suffer from random disorder effects. It is thus an important issue to know the influence of disorder on the device characteristic. While the position of the holes can be controlled to a high accuracy,

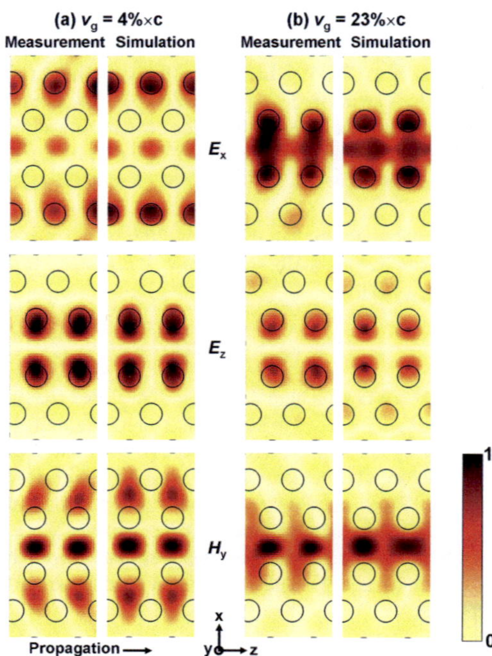

Fig. 4.22. Field distribution at frequencies with low (a) and high (b) group velocity. Excellent agreement between measured and simulated field amplitudes is observed. For low group velocity, the fields extend considerably into the PC region, whereas for high group velocity, the fields are mainly confined to the waveguide core.

the radii and shape of the holes might vary. In the microwave model, it is possible to deterministically introduce disorder, as the equivalent optical structure accuracy is about 0.5 nm. In this work, we study the influence of radial disorder on the PC-WG.

We chose a normally distributed disorder with a standard deviation of $\sigma = 0.05 \times r_o$, where r_o is the nominal radius of the holes. We fabricated three different samples of 15 periods of the PC-WG and conducted CW measurements. The group velocity behavior is displayed in Fig. 4.23(a). For the disordered samples, the traces oscillate around the trace of the ideal sample, while the ensemble average converges to it. So broadband propagation at low group velocity is still possible even with disorder.

In Fig. 4.23(b) it can be seen that the transmission through the disordered samples may increase or decrease compared to the case without disorder, depending on the exact disorder realization. However, the ensemble average of the transmission will be lower than for the case without disorder.

Peaks and dips in the group velocity and the transmission behavior can be observed in Fig. 4.23. The disorder generates random resonances, where the characteristics depend on the

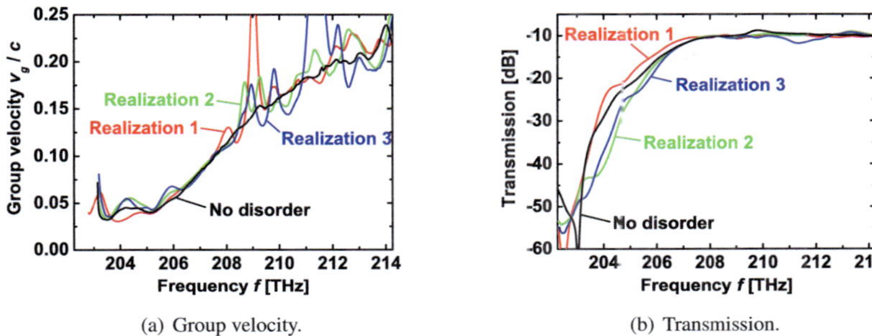

(a) Group velocity.   (b) Transmission.

Fig. 4.23. (a) Group velocity and (b) transmission from measurement of disordered microwave model samples with a length of 15 periods. The radial disorder is normally distributed with a standard deviation of 5% of the nominal radius. The curves are scaled to optical frequencies and slightly shifted towards each other to account for the permittivity tolerance in various samples. (a) The group velocity traces oscillate around the trace of the ideal sample, and broadband transmission at low group velocity is still possible. (b) Depending on the exact realization, transmission may increase or decrease. The ensemble average is lower than for the case without disorder.

exact disorder realization. For long samples, even photon localization might occur (Anderson localization [153]).

To test how accurately the disordered PC can be described by simulations, we also conducted a FIT transient simulation of realization 1, the results are displayed in Fig. 4.24. The root mean square (RMS) of the relative differences between the measurement and the simulation is around 5% for the group velocity and around 15% for the linear transmission over a 10 THz bandwidth, which gives a high credibility also to the simulation of disorder. For the

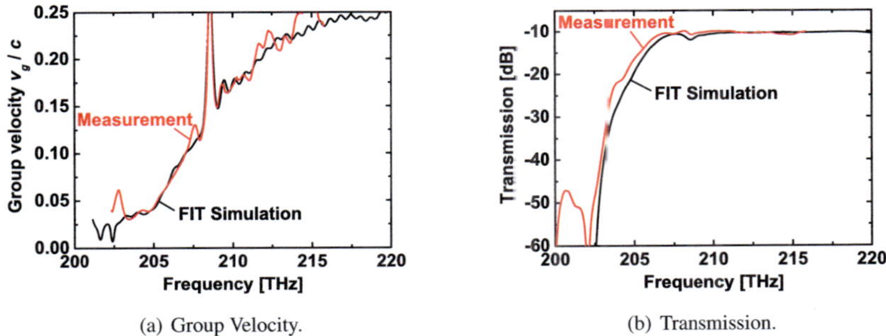

(a) Group Velocity.   (b) Transmission.

Fig. 4.24. Comparison of measurement results to FIT simulations for a disordered structure. The microwave experiments are very well predicted by the simulations, with RMS deviations between experiment and simulation of 5% for the group velocity and 13% for the transmission in linear scale.

device without disorder, the RMS differences between experiment and simulation are 5 % for the group velocity, Fig. 4.20(a), and 13 % for the linear transmission, Fig. 4.19(a).

If the material losses are turned off for the simulations, which resembles the situation for optical devices, transmission as well as reflection increase, but also the resonances caused by disorder become narrower.

4.2.3 Conclusion

We present a guided-wave microwave model for the experimental proof-of-concept for novel photonic devices with high index contrast. We exploit the high fabrication accuracy of the enlarged structures, and the availability of precise microwave measurement equipment near 10 GHz. A broadband slow-light PC-WG is realized in a ceramic-filled plastic that exhibits a region with a group velocity of 3.4 % of the vacuum speed of light in a 1.2 THz (optical) bandwidth and a region with an (optical) chromatic dispersion of 3.9 ps / (mm nm) in a 1.1 THz bandwidth. Based on these experiments, we verify that FIT simulations predict the behavior of PC-devices correctly. We show that a 5 % disorder in the hole radii does not significantly impair the group velocity characteristics thus proving the feasibility of a realistic device.

4.3 Verification of Numerics

The first step in the process to build functional integrated PC devices is the numerical design of the PC-WG, which is done with band calculations, as presented in Section 2.1. Various numerical methods are available and commonly used to achieve the design, yet not all are equally feasible, and considerable differences in simulation time and hardware requirements exist. After the design of the PC-WGs, also complete devices need to be simulated to gain better insights into the device behavior and to avoid unneeded fabrication runs. In Section 4.2, it has also been verified that the FIT method can well predict the behavior a PC-WG device including coupling sections and the transition from strip-WG to PC-WG, even in the presence of systematically introduced radial disorder. The microwave material, however, exhibits significant material losses, which are not present in silicon or glass, and the behavior especially of disordered structures might become different if the material losses become negligible.

In this section, we evaluate several methods for band calculations, which are the FIT, FDTD, PWE and GME methods. As has been already presented in Section 4.2, the FIT method accurately predicts the measured group velocity and chromatic dispersion behavior of a scaled PC-WG in the microwave regime. We find that the GME method is a very fast design tool that predicts the PC-WG behavior correctly. Furthermore, we study the feasibility of the FIT, GME and FDTD methods to determine the influence of disorder on the silicon device characteristics, as needed for the work presented in Section 2.2. We find that the results from FIT and FDTD are trustworthy, whereas the GME method needs to be used more carefully.

This section is organized as follows: In Subsection 4.3.1, different design tools are benchmarked, and in Subsection 4.3.2, the numerics verification to study disordered PC-WG is presented.

The contents of Subsection 4.3.1 have been published as part of a journal article, [2] or [J2] on page 165.

4.3.1 Benchmark of Different Design Tools

A PC-WG with a certain group velocity or dispersion characteristic is designed with numerical band calculations. To choose among the different numerical methods, accuracy, calculation speed and memory requirements have to be evaluated. For the slow-light PC-WG, we benchmark the four different methods mentioned in Appendix A.2, namely the FIT eigenmode solver, FDTD, PWE and GME methods. Fig. 4.25 shows the group velocity curves obtained from the band diagrams. The curves from FIT eigenmode, PWE and GME solver are almost identical and predict the behavior of the component correctly, as verified by microwave experiments, Fig. 4.20(a). The FDTD curve, however, is shifted in frequency and deviates more from the other results.

In Table 4.2, the simulation runtime for 20 wavevector points of the band diagram and the memory usage of the different methods are compared. For the simulations, a discretization of around 20 cells per period and symmetric boundary conditions were used. The FIT eigenmode, FDTD and GME simulations were run on a standard computer, whereas the PWE simulations were run on 12 nodes of a unix cluster, as the memory size of a single computer was not sufficient.

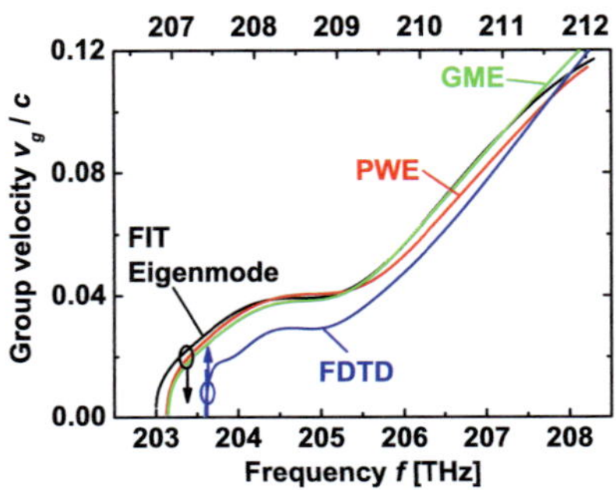

Fig. 4.25. Group velocity obtained from different methods to calculate the band diagram. The FIT eigenmode, PWE and GME methods predict the correct behavior. The upper frequency scale is for the FDTD-result.

Obviously, the GME method is fastest, and also the FIT eigenmode method has a reasonable runtime. However, the FDTD and PWE simulations are much slower, and the FDTD results are less reliable for the chosen discretization and simulation time.

	FIT Eigenmode	FDTD	PWE (12 CPU)	GME
Runtime (Minutes)	50	1000	360	5
Memory (MB)	30	35	> 570	115

Table 4.2. Runtime and memory usage for different methods to calculate band diagrams.

4.3.2   Simulation of Disorder-Induced Losses

The numerical determination of disorder-induced losses in PC-WGs is a non-trivial task. With full-numerical methods like FIT or FDTD it is in principle possible to study arbitrary geometries without neglecting any effects, however the simulations become very time- and memory-consuming, which limits the maximum structure size, maximum simulation time and the minimum discretization step that can be used. The disorder causes only a small deviation to the nominal structure, which in the case of radial disorder is a small change in the PC hole radii around their nominal values. These deviations need to be resolved by the discretization, and as the disorder is statistical, a sufficiently large number of disorder realizations needs to be treated. In contrast to the full-numerical methods, semi-analytical methods like the GME method are much faster and discretization issues play a smaller role. However, the semi-analytical methods rely on physical models that do not consider all possible effects and in many cases use perturbative approximations. The ultimate proof could be given by experiments, however the fabrication imperfections are always present and cannot be controlled, only a limited number of physical quantities can be directly measured, and measurement equipment has a limited bandwidth and accuracy.

To gain trust in the numerically determined losses, we compare the results of independent simulation methods, which are FIT, FDTD and GME. In the GME method, the eigenmodes of a unit cell of the PC-WG are determined and the out-of-plane and backscattering losses are obtained from a perturbative calculation, which means that mode coupling is not considered. In the FIT and FDTD simulations, however, the slow-light PC mode needs to be efficiently excited by a strip-WG, which is in our case achieved by butt-coupling the strip-WG to the PC-WG and gradually slowing down the PC mode in the first PC periods by a taper. The influence of the coupling section and the taper needs to be removed to obtain the transmission and reflection caused by the disordered section only, which is achieved with the TRL calibration technique, similarly as in the microwave experiments.

Numerical Calibration Procedure

To apply the TRL calibration numerically to the simulation results, three reference simulations are needed in addition to the disorder-simulations, see Appendix A.4 for more details. The reference structures are a thru, which is the direct connection of the PC tapers fed by the strip-WGs without a section of disordered slow-light PC in between, a reflect, realized by a reflecting metal plate at the end of the taper, and a line, which is a section of a non-reflecting ideal PC between the tapers. The three calibration structures are shown in Fig. 4.26, for the 'Air' structure of Section 2.2.

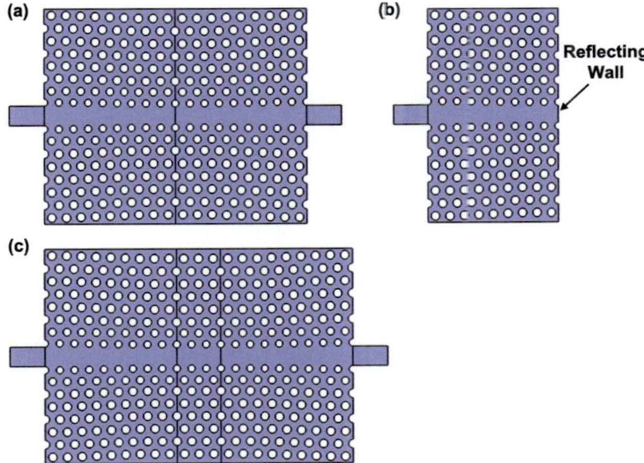

Fig. 4.26. TRL calibration structures for disorder simulations. (a) Through structure, where the PC tapers are directly connected. (b) Reflect structure with an ideally reflecting metal plate at the taper end. (c) Line structure with 3 periods of PC-WG between the two PC tapers.

As a result of the calibration procedure, the transmission and reflection characteristics of the disordered PC section can be calculated, and the group velocity can be inferred from the phase information of the transmission. To obtain the ensemble average, the transmission and reflection results of the different disorder realizations are averaged.

Comparison of Ensemble Averaged Results

We have chosen the broadband slow-light PC-WG of the scaled microwave experiments, Subsection 2.2.3, as the reference for the numerics comparison. The W0.75 line-defect WG has holes with equal radii $r/a = 0.25$ and is inside a membrane of thickness $h/a = 0.6$ with permittivity $\varepsilon_r = 10$, where the PC period is $a = 0.45\,\mu$m.

First, the results from the two full-numerical methods FIT and FDTD are compared, where the same 10 disorder realizations with a length of $15\,a$ are simulated with a disorder RMS of $\sigma = 1\% \times r = 1.125$ nm. The simulation domain was reduced to $1/4$ of the original structure size by exploiting two symmetry conditions. The first condition comes from the vertical symmetry; the structure is symmetric with respect to the plane that bisects the high-index slab. This symmetry decouples the TE and TM modes. The second condition is introduced by setting the hole radii on both sides of the WG core to the same values, which introduces a symmetry plane that is vertical to the slab and along the WG direction. This symmetry plane decouples the even and odd modes and does not represent the real situation, but is sufficient for the numerics verification. The averaged calibrated transmission and reflection is shown in Fig. 4.27. A constant frequency shift is introduced between the results of the two methods such that the group velocity characteristics overlap. It can be observed that the results of the two methods agree very well, although the two methods use different algorithms and different discretization. The agreement verifies the suitability of the methods. The oscillations in the FDTD curves come from the finite simulation time. The FIT method is faster and needs less memory than the FDTD method, which makes the FIT method the better choice for the disorder simulations.

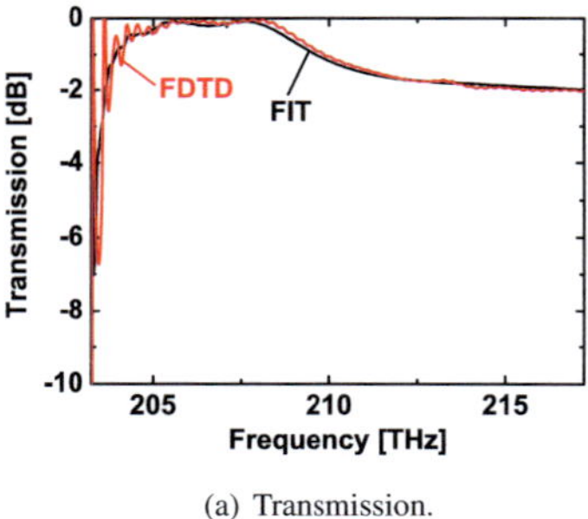

(a) Transmission.

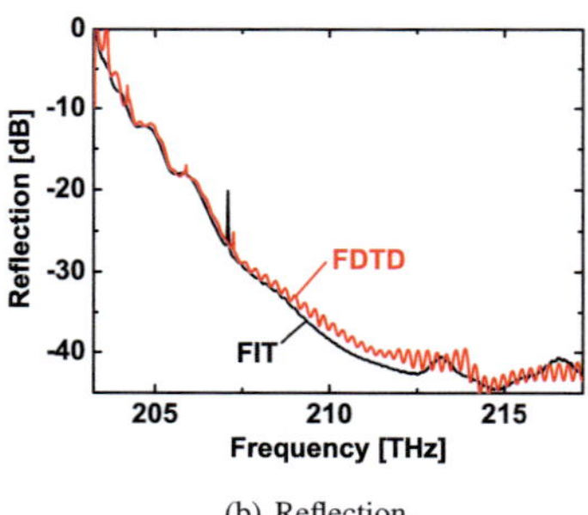

(b) Reflection.

Fig. 4.27. Comparison of FIT and FDTD results for disordered structures. The transmission and reflection characteristics from FIT and FDTD agree nicely, which verifies the results. The oscillations in the FDTD results are caused by the finite simulation time. All curves represent an average over 10 disorder realizations.

In the next step, the results from FIT are compared to the results from GME for the same structure and disorder level. The FIT results are again calibrated and averaged, and in addition the transmission normalized to the transmission of an ideal PC-WG without disorder in order

to only reflect the excess losses. The number of simulated disorder realizations is 110 for
the GME method, but only 10 for the FIT due to the long simulation times. The results are
shown in Fig. 4.28. We find that the FIT results for the structure being symmetric with respect
to the plane vertical to the slab and along the WG direction deviate considerably from the
GME curves. To find out how this somewhat unphysical symmetry condition influences the
results, we conduct simulations where this symmetry is removed; the results are also plotted in
Fig. 4.28. Indeed, the results change considerably, which reveals that the use of the symmetry is
not allowed. However, even without symmetry in the FIT simulations, the results deviate from
the GME results, especially in the reflection. Although the general trends are the same from
both methods, the exact results are different.

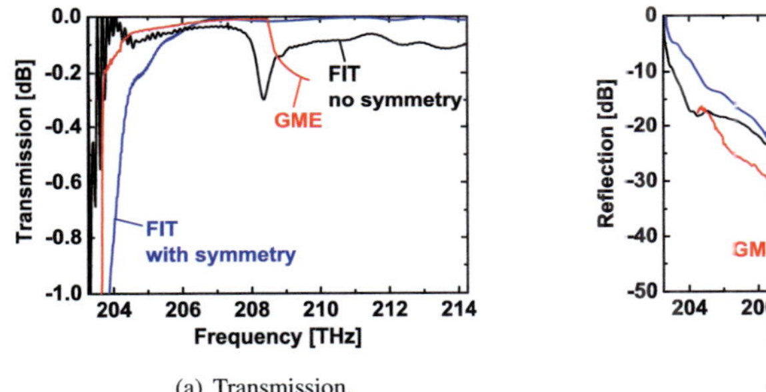

(a) Transmission.　　　　(b) Reflection.

Fig. 4.28. Comparison of FIT and GME results for disordered structures. (a) The FIT results without structure
symmetry generally agree with the GME results, but the FIT results with structure symmetry deviate considerably. (b) The FIT and GME results do not agree, however are more similar for FIT simulations without structure
symmetry.

Comparison of Single Realization

As an additional step, we also compare the results from FIT and FDTD for one disorder realization of the 'Air' structure of Section 2.2. The symmetry plane that is vertical to the slab and
along the WG direction has been removed. As can be seen in Fig. 4.29, the agreement is in general good, however some features in the spectra are at different frequencies. This is most likely
caused by the different discretizations used in both methods; the FIT method uses a nonuniform
mesh with a larger minimum mesh size than the uniform mesh in the FDTD method.

4.3.3 Conclusion

We evaluate and compare different numerical tools for PC-WG band calculations used to design
the group velocity and dispersion characteristics. Based on previous microwave experiments,
we prove that the GME, FIT and PWE methods predict the behavior correctly, but that GME is
the fastest tool, followed by FIT. We also examine the suitability of different numerical methods
to model disorder effects in PC-WGs. We find that results from FIT agree well with results from

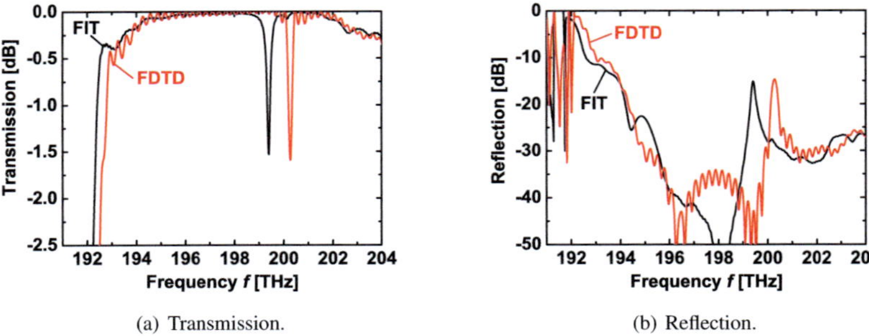

(a) Transmission.   (b) Reflection.

Fig. 4.29. Comparison of FIT and FDTD results for a disordered 'Air' structure. The FIT and FDTD curves agree in general, however some features are at different spectral positions, most likely caused by a different discretization. The oscillations in the FDTD results come from the finite simulation time.

the FDTD method, which gives high confidence in the reliability of FIT for disorder simulations. The much faster GME method, however, can only be used for qualitative or comparative predictions, as the results deviate more from those of the other methods.

Appendix

A.1 Mathematical Transformations and Signal Representation

In this section, we briefly state the mathematical transformations and the signal representation used in the thesis.

A.1.1 Fourier Transform

A signal or linear system response can be represented in time domain as a function of time $x(t)$, or in frequency domain as a function of frequency $X(f)$. The transformation from time-domain into frequency-domain is accomplished with the Fourier transform, and the transformation from frequency-domain into time-domain with the inverse Fourier transform. The relations are:

$$X(f) = \int_{-\infty}^{+\infty} x(t)\,e^{-j2\pi ft}\,dt, \tag{A.1}$$

$$x(t) = \int_{-\infty}^{+\infty} X(f)\,e^{j2\pi ft}\,df. \tag{A.2}$$

In addition to the time-frequency Fourier transform, an analog three-dimensional spatial Fourier transform can be defined that transforms a function dependent on spatial coordinates into the wavevector space, where the function is expanded into plane waves, and vice versa.

A.1.2 Signal Representation

Throughout this work, we represent optical signals and fields as complex quantities. The time-dependence of the carrier is of the form $e^{j\omega t}$, and the propagation along a WG in $z-$direction can be expressed as $e^{-j\beta z}$. The measurable physical quantities of the fields correspond to the real part of the complex representation. The complex functions are analytical, and the imaginary part of the function is obtained as the Hilbert transform of the real part. If the analytical function is represented in the frequency domain by using the Fourier transform, only spectral components at positive frequencies exist, whereas the Fourier transform of the corresponding real function has components at positive and negative frequencies, and the components at negative frequencies are the complex conjugate values of the components at positive frequencies mirrored at the origin.

The reponse $y(t)$ of a signal $x(t)$ transmitted through a linear time-invariant system with impulse response $g(t)$ can be calculated as the convolution of the input signal and the impulse response:

$$y(t) = x(t) * g(t) = \int_{-\infty}^{+\infty} x(\tau)\, g(t - \tau)\, d\tau. \tag{A.3}$$

In frequency domain, the spectrum $Y(f)$ of the output signal $y(t)$ can be obtained by multiplying the spectrum $X(f)$ of the input signal $x(t)$ with the Fourier transform $G(f)$ of the impulse response $g(t)$:

$$Y(f) = X(f)\, G(f). \tag{A.4}$$

The function $G(f)$ is also called the transfer function of the system. The properties of a linear optical device can be characterized with its $S-$parameters as a function of frequency, Eq. (A.24), and the device transfer functions for the transmitted or reflected signals are just the corresponding $S-$parameters, $S_{21}(f)$ or $S_{11}(f)$, respectively.

If a system is excited with a harmonic $x(t) = A \exp(j 2\pi f_0 t)$ of complex amplitude A and frequency f_0, the output signal is just the input signal multiplied with the transfer function at frequency f_0, $y(t) = G(f_0)\, A \exp(j 2\pi f_0 t)$.

A.1.3 Laplace Transform

The Laplace transform maps a causal function $x(t)$ into the complex $s-$space. The transformation equations are:

$$X(s) = \int_0^{+\infty} y(t)\, e^{-st}\, dt, \tag{A.5}$$

$$x(t) = \frac{1}{j 2\pi} \int_{c-j\infty}^{c+j\infty} Y(s)\, e^{st}\, ds. \tag{A.6}$$

The Laplace transform can e. g. be used to solve differential equations, where correspondence tables can be consulted to transform the functions. Correpondence pairs are denoted in the form $x(t) \circ\!\!-\!\!\bullet X(s)$. In Appendix A.6, we use the Laplace transform to solve the differential equations of the loss model.

A.2 Numerical Modeling Tools

Finite Integration Technique (FIT)

Maxwell's equations are solved in integral form, and the simulation domain is discretized in a rectangular grid. The eigenmode solver works in the frequency domain, and finds the eigenmodes and eigenfrequencies of a structure. The transient solver operates in the time domain and uses the leapfrog algorithm to compute the field evolution inside the structure at discrete time steps. We use the commercially available software package CST Microwave Studio [160] to calculate band diagrams with the eigenmode solver, and we simulate transmission and reflection of complete devices with the transient solver.

Finite Difference Time Domain (FDTD) Method

Maxwell's equations are solved in differential form, and the simulation domain is discretized in a uniform grid. Again, the leapfrog algorithm is applied to compute the field evolution inside the structure at discrete time steps. With the commercially available software RSoft FullWAVE [161], we generate band diagrams, and we simulate transmission and reflection of a structure.

Plane Wave Expansion (PWE) Method

The domain is periodically continued in all three space directions, and the fields are expanded into plane waves. The eigenmodes and eigenfrequencies of a structure are found in the frequency domain. We use the freely available package MIT Photonic-Bands (MPB) [162] to calculate band diagrams.

Guided Mode Expansion (GME) Method

The technique is suited for slab structures, and the fields are expanded into guided modes of an effective slab. In the plane orthogonal to the slab, the expansion is into plane waves. This is again a frequency domain technique, where the eigenmodes and eigenfrequencies of a structure are found. Using the freely available program GME [163] we generate band diagrams and calculate losses in a disordered structure.

A.3 Mode Orthogonality and Group Velocity

The orthogonality relation for modes of a PC-WG, Eq. (1.17), can be derived from the Lorentz reciprocity theorem, which will be shown first following [164]. We pick two harmonic solutions $\mathbf{E}_{p,q}$, $\mathbf{H}_{p,q}$ of Maxwell's equations Eqs. (1.9)-(1.10) at frequencies $\omega_{p,q}$, which satisfy the following equations:

$$\nabla \times \mathbf{E}_{p,q}(\mathbf{r}) = -j\,\omega_{p,q}\,\mu_0\,\mathbf{H}_{p,q}(\mathbf{r})\,, \tag{A.7}$$

$$\nabla \times \mathbf{H}_{p,q}(\mathbf{r}) = j\,\omega_{p,q}\,\varepsilon_0\varepsilon_r\,\mathbf{E}_{p,q}(\mathbf{r})\,. \tag{A.8}$$

For two vector fields $\mathbf{A}, \mathbf{B}$, the identity $\nabla \cdot (\mathbf{A} \times \mathbf{B}) = \mathbf{B} \cdot (\nabla \times \mathbf{A}) - \mathbf{A} \cdot (\nabla \times \mathbf{B})$ holds, which can be used to calculate the following result:

$$\nabla \cdot \left(\mathbf{E}_q \times \mathbf{H}_p^* - \mathbf{H}_q \times \mathbf{E}_p^*\right) = j\left(\omega_p - \omega_q\right)\left[\mu_0 \mathbf{H}_p^* \cdot \mathbf{H}_q + \varepsilon_0\varepsilon_r \mathbf{E}_p^* \cdot \mathbf{E}_q\right]. \tag{A.9}$$

This equation is integrated in a volume V (bound by the closed surface F with normal vector $\mathbf{F}$), and the Gauss' theorem $\int_V \nabla \cdot \mathbf{A}\, dV = \oint_F \mathbf{A}\, d\mathbf{F}$ is applied to the left hand side to yield the *Lorentz reciprocity theorem*:

$$\oint_F \left(\mathbf{E}_q \times \mathbf{H}_p^* - \mathbf{H}_q \times \mathbf{E}_p^*\right) d\mathbf{F} = j\left(\omega_p - \omega_q\right) \int_V \left(\mu_0 \mathbf{H}_p^* \cdot \mathbf{H}_q + \varepsilon_0\varepsilon_r \mathbf{E}_p^* \cdot \mathbf{E}_q\right) dV\,. \tag{A.10}$$

If the two modes have the same frequency $\omega_p = \omega_q$, the right hand side of Eq. (A.10) vanishes. We consider bound modes that propagate in the waveguide along z-direction and are confined to the core. The volume V is chosen in a z-range from z_1 to z_2, $z_1 < z_2$, and the extension in x and y-direction should be large enough so that the fields have decayed at the boundaries far away from the core, and the surface integral on the left hand side of Eq. (A.10) needs only be taken at the planes $z = z_2$ and $z = z_1$:

$$\begin{aligned}
0 &= \oint_F \left(\mathbf{E}_q \times \mathbf{H}_p^* - \mathbf{H}_q \times \mathbf{E}_p^*\right) d\mathbf{F} \\
&= \int_{z=z_2} \left(\mathbf{E}_q \times \mathbf{H}_p^* - \mathbf{H}_q \times \mathbf{E}_p^*\right) \cdot e_z\, dA - \int_{z=z_1} \left(\mathbf{E}_q \times \mathbf{H}_p^* - \mathbf{H}_q \times \mathbf{E}_p^*\right) \cdot e_z\, dA\,.
\end{aligned} \tag{A.11}$$

For the two modes p, q in the PC-WG, the $\mathbf{E}$ and $\mathbf{H}$-fields can be expressed using the Bloch theorem Eq. (1.20) with the mode field amplitude $\hat{\mathbf{E}}_{p,q}$, $\hat{\mathbf{H}}_{p,q}$:

$$\mathbf{H}_{p,q}(x,y,z) = e^{-jk_{p,q}z}\,\hat{\mathbf{H}}_{p,q}(x,y,z) \quad \text{with } \hat{\mathbf{H}}_{p,q}(x,y,z) = \hat{\mathbf{H}}_{p,q}(x,y,z+a)\,, \tag{A.12}$$

$$\mathbf{E}_{p,q}(x,y,z) = e^{-jk_{p,q}z}\,\hat{\mathbf{E}}_{p,q}(x,y,z) \quad \text{with } \hat{\mathbf{E}}_{p,q}(x,y,z) = \hat{\mathbf{E}}_{p,q}(x,y,z+a)\,. \tag{A.13}$$

We now substitute these modes into Eq. (A.11) and choose the z-positions $z_1 = z$, $z_2 = z+a$ with arbitrary z to obtain:

$$\begin{aligned}
0 &= \int_z \left(\mathbf{E}_q \times \mathbf{H}_p^* - \mathbf{H}_q \times \mathbf{E}_p^*\right) \cdot e_z\, dA - \int_{z-a} \left(\mathbf{E}_q \times \mathbf{H}_p^* - \mathbf{H}_q \times \mathbf{E}_p^*\right) \cdot e_z\, dA \\
&= e^{-j(k_q-k_p)z}\left(e^{-j(k_q-k_p)a} - 1\right) \int_z \left(\hat{\mathbf{E}}_q \times \hat{\mathbf{H}}_p^* - \hat{\mathbf{H}}_q \times \hat{\mathbf{E}}_p^*\right) \cdot e_z\, dA\,.
\end{aligned} \tag{A.14}$$

In the second line of this equation, either the factor $\left(\mathrm{e}^{-\mathrm{j}\left(k_q-k_p\right)a}-1\right)$ or the integral expression $\int_z \left(\hat{\mathbf{E}}_q \times \hat{\mathbf{H}}_p^* - \hat{\mathbf{H}}_q \times \hat{\mathbf{E}}_p^*\right)\cdot e_z\,\mathrm{d}A$ has to vanish for the equation to hold. This leads to the *mode orthogonality relation* for PC-WGs ($k_{p,q}$ is inside the first Brillouin zone $-\pi/a\dots\pi/a$):

$$\int_z \left(\hat{\mathbf{E}}_q \times \hat{\mathbf{H}}_p^* - \hat{\mathbf{H}}_q \times \hat{\mathbf{E}}_p^*\right)\cdot e_z\,\mathrm{d}A = 4\,\delta_{p,q}\,P_p\,. \tag{A.15}$$

Here, $\delta_{p,q}$ is the Kronecker delta, and P_p is the cross-section power of mode p. For $p = q$, the power of the mode is obtained

$$P_p = \frac{1}{4}\int_z \left(\hat{\mathbf{E}}_q \times \hat{\mathbf{H}}_q^* - \hat{\mathbf{H}}_q \times \hat{\mathbf{E}}_q^*\right)\cdot e_z\,\mathrm{d}A = \frac{1}{2}\int_z \mathfrak{R}\left(\hat{\mathbf{E}}_q \times \hat{\mathbf{H}}_q^*\right)\cdot e_z\,\mathrm{d}A\,, \tag{A.16}$$

where the conversion on the right hand side was achieved using the identities $-\hat{\mathbf{H}}_q \times \hat{\mathbf{E}}_q^* = \hat{\mathbf{E}}_q^* \times \hat{\mathbf{H}}_q$ and $w + w^* = 2\mathfrak{R}\left(w\right)$ for a complex number w. On the right-hand side of Eq. (A.16), the Poynting vector $\mathbf{S} = \frac{1}{2}\hat{\mathbf{E}} \times \hat{\mathbf{H}}^*$ appears, and the real part of its z-component specifies the power flow in propagation direction. If integrated over a cross-section area, the total power is obtained, which is constant along z if no losses occur.

From the mode field distribution, it is also possible to infer the group velocity $v_g = \frac{\mathrm{d}\omega}{\mathrm{d}k}$ of the mode. This can be derived starting from Lorentz reciprocity theorem, Eq. (A.10), but with initially different frequencies for the two modes p and q [164]. Then, the same steps as above are followed and the PC-WG modes Eqs. (A.12)-(A.13) substituted, which results in a more general relation of Eq. (A.14) with a non-vanishing left hand side (the position z can without loss of generality be chosen as 0 by just shifting the structure):

$$\mathrm{j}\left(\omega_p - \omega_q\right)\int_V \left(\mu_0\hat{\mathbf{H}}_p^* \cdot \hat{\mathbf{H}}_q + \varepsilon_0\varepsilon_r\hat{\mathbf{E}}_p^* \cdot \hat{\mathbf{E}}_q\right)\mathrm{d}V$$
$$= \left(\mathrm{e}^{-\mathrm{j}\left(k_q-k_p\right)a}-1\right)\int_z \left(\hat{\mathbf{E}}_q \times \hat{\mathbf{H}}_p^* - \hat{\mathbf{H}}_q \times \hat{\mathbf{E}}_p^*\right)\cdot e_z\,\mathrm{d}A\,. \tag{A.17}$$

If the same mode is chosen for p and q and the frequency ω_q approaches ω_q, the mode fields become the same, $\hat{\mathbf{E}}_p = \hat{\mathbf{E}}_q = \hat{\mathbf{E}}$, $\hat{\mathbf{H}}_p = \hat{\mathbf{H}}_q = \hat{\mathbf{H}}$, and a differential is obtained (where $\mathrm{e}^{-\mathrm{j}\left(k_q-k_p\right)a}$ becomes 1):

$$-\mathrm{j}\,\mathrm{d}\omega\int_V \left(\mu_0\left|\hat{\mathbf{H}}\right|^2 + \varepsilon_0\varepsilon_r\left|\hat{\mathbf{E}}\right|^2\right)\mathrm{d}V = -\mathrm{j}\,a\,\mathrm{d}k\int_z \left(\hat{\mathbf{E}} \times \hat{\mathbf{H}}^* - \hat{\mathbf{H}} \times \hat{\mathbf{E}}^*\right)\cdot e_z\,\mathrm{d}A\,. \tag{A.18}$$

From this equation, the *group velocity* can be inferred:

$$v_g = \frac{\mathrm{d}\omega}{\mathrm{d}k} = \frac{a\int_z \left(\hat{\mathbf{E}} \times \hat{\mathbf{H}}^* - \hat{\mathbf{H}} \times \hat{\mathbf{E}}^*\right)\cdot e_z\,\mathrm{d}A}{\int_V \left(\mu_0\left|\hat{\mathbf{H}}\right|^2 + \varepsilon_0\varepsilon_r\left|\hat{\mathbf{E}}\right|^2\right)\mathrm{d}V} = \frac{2a\int_z \mathfrak{R}\left(\hat{\mathbf{E}} \times \hat{\mathbf{H}}^*\right)\cdot e_z\,\mathrm{d}A}{\int_V \left(\mu_0\left|\hat{\mathbf{H}}\right|^2 + \varepsilon_0\varepsilon_r\left|\hat{\mathbf{E}}\right|^2\right)\mathrm{d}V}\,. \tag{A.19}$$

The surface integral in the numerator is performed over a cross-section orthogonal to propagation direction, and the volume integration in the denominator is over a unit cell of the PC of length a. The integrand in the numerator is the real part of the Poynting vector in $z-$direction, and the integral quantifies the total power flow of the mode. In the denominator, the integrand is the sum of the magnetic and electric energy density in the unit cell. Thus, for a fixed amount of transmitted power, the stored energy increases inversely proportional to the group velocity of a mode.

It can be shown that the integral of the electric and magnetic energy density over a unit cell are same, which will be proven in the following. With the same procedure that led to Eq. (A.9), it can be shown that a solution of Maxwell's equations satisfies

$$\nabla \cdot (\mathbf{E} \times \mathbf{H}^*) = j\omega \left(\varepsilon_0 \varepsilon_r |\mathbf{E}|^2 - \mu_0 |\mathbf{H}|^2 \right) . \tag{A.20}$$

The Gauss' theorem (as stated after Eq. (A.9) can be applied to this equation, and results in

$$j\omega \int_V \left(\varepsilon_0 \varepsilon_r |\mathbf{E}|^2 - \mu_0 |\mathbf{H}|^2 \right) dV = \oint_F (\mathbf{E} \times \mathbf{H}^*) \, d\mathbf{F} . \tag{A.21}$$

If a PC mode according to Eqs. (A.12)-(A.13) is substituted in Eq. (A.21) and the volume V is chosen as a PC unit cell from $z_1 = z$ to $z_2 = z + a$, the surface integral over the closed surface F can be represented as a sum of two surface integrals at the constant z-planes at z and $z + a$:

$$\begin{aligned} j\omega \int_V \left(\varepsilon_0 \varepsilon_r |\hat{\mathbf{E}}|^2 - \mu_0 |\hat{\mathbf{H}}|^2 \right) dV &= \int_{z+a} (\mathbf{E} \times \mathbf{H}^*) \cdot e_z \, dA - \int_z (\mathbf{E} \times \mathbf{H}^*) \cdot e_z \, dA \\ &= \int_{z+a} (\hat{\mathbf{E}} \times \hat{\mathbf{H}}^*) \cdot e_z \, dA - \int_z (\hat{\mathbf{E}} \times \hat{\mathbf{H}}^*) \cdot e_z \, dA \\ &= 0, \end{aligned} \tag{A.22}$$

which results in the *equality of the electric and magnetic energy density integrated in a unit cell* V:

$$\int_V \varepsilon_0 \varepsilon_r |\hat{\mathbf{E}}|^2 \, dV = \int_V \mu_0 |\hat{\mathbf{H}}|^2 \, dV . \tag{A.23}$$

This equality can be used in Eq. (A.19) to simplify the calculation in the denominator.

A.4 Through-Reflect-Line (TRL) Calibration

The Through-Reflect-Line (TRL) calibration is a method than can mathematically remove the influence of a parasitic device section from the measurement or simulation results [165]. The waveguide-based linear optical device is modeled as a two-port network, where the WGs in each port are single-moded or higher-order modes are irrelevant. The behavior of the network is characterized in frequency domain, and the fields in the port planes are expanded in the forward and backward propagating WG modes. The complex amplitudes of the modes with propagation direction into the network are denoted as $a_{1,2}$, and of the modes with propagation direction out of the network as $b_{1,2}$, where the index stands for the port number. This is the concept of S-parameters, and the scattering-matrix $\mathbf{S}$ characterizes the network and relates the outgoing amplitudes with the ingoing amplitudes (see also Fig. A.4 and Eq. (A.89) according to:

$$\begin{pmatrix} b_1 \\ b_2 \end{pmatrix} = \begin{pmatrix} S_{11} & S_{12} \\ S_{21} & S_{22} \end{pmatrix} \begin{pmatrix} a_1 \\ a_2 \end{pmatrix}. \tag{A.24}$$

The actual device under test (DUT) is surrounded by two other networks, denoted as A and B, e. g. the coupling structures at the input and output of the device, but only the characteristics of the concatenated device can be simulated or measured, see Fig. A.1(a). With three additional calibration simulations or measurements, where the DUT is replaced by a direct connection of A and B ('thru' or 'through'), by a reflection ('reflect') or by a non-reflecting WG-section ('line'), the influence of networks A and B can be removed from the result by a mathematical calibration procedure, see Fig. A.1(b). This procedure is explained in the following. First, we

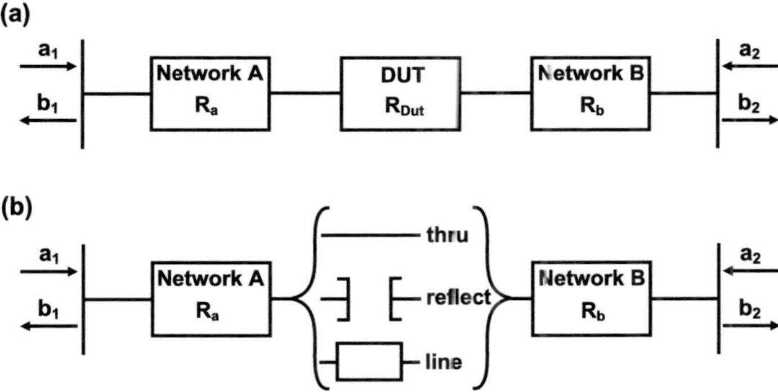

Fig. A.1. TRL calibration scheme. (a) The device under test (DUT) is surrounded by two networks A and B. (b) Three additional calibration measurements or simulations allow to mathematically remove the influence of networks A and B. The standards are a 'thru' standard, which is the direct connection of A and B, a 'reflect' standard, which is a highly reflecting termination, and a 'line' standard, which is a section of a non-reflecting waveguide.

need to introduce the wave cascading matrix or **R**-matrix, which can be calculated from the scattering matrix **S** in the following way:

$$\begin{pmatrix} b_1 \\ a_1 \end{pmatrix} = \frac{1}{S_{21}} \begin{pmatrix} -\Delta & S_{11} \\ -S_{22} & 1 \end{pmatrix} \begin{pmatrix} a_2 \\ b_2 \end{pmatrix} = \mathbf{R} \begin{pmatrix} a_2 \\ b_2 \end{pmatrix},$$
$$\Delta = S_{11}S_{22} - S_{12}S_{21}.$$

(A.25)

The **R**-matrices are chosen because a concatenation of networks results in the multiplication of the corresponding **R**-matrices. The thru-connection can thus be expressed as

$$\mathbf{R}_t = \mathbf{R}_a \mathbf{R}_b,$$

(A.26)

and the line connection as

$$\mathbf{R}_d = \mathbf{R}_a \mathbf{R}_l \mathbf{R}_b,$$

(A.27)

with $\mathbf{R}_l = \begin{pmatrix} e^{-jkl} & 0 \\ 0 & e^{jkl} \end{pmatrix}$ being the **R**-matrix of an ideal nonreflection line with (complex) propagation constant k and length l. The matrices $\mathbf{R}_t$ and $\mathbf{R}_d$ are known from the calibration simulations or measurements. Solving Eq. (A.26) for $\mathbf{R}_b$ and inserting into (A.27) eliminates $\mathbf{R}_b$ and yields

$$\mathbf{T}\mathbf{R}_a = \mathbf{R}_a \mathbf{R}_l \quad \text{with} \quad \mathbf{T} = \mathbf{R}_d \mathbf{R}_t^{-1}.$$

(A.28)

This can be rewritten inserting $\mathbf{T} = \begin{pmatrix} t_{11} & t_{12} \\ t_{21} & t_{22} \end{pmatrix}$ and $\mathbf{R}_a = \begin{pmatrix} r_{11} & r_{12} \\ r_{21} & r_{22} \end{pmatrix}$ as

$$\begin{pmatrix} t_{11} & t_{12} \\ t_{21} & t_{22} \end{pmatrix} \begin{pmatrix} r_{11} & r_{12} \\ r_{21} & r_{22} \end{pmatrix} = \begin{pmatrix} r_{11} & r_{12} \\ r_{21} & r_{22} \end{pmatrix} \begin{pmatrix} e^{-jkl} & 0 \\ 0 & e^{jkl} \end{pmatrix}.$$

(A.29)

This matrix equation contains four equations, but it can be shown that only three are linearly independent, which means that only three parameters can be determined. Manipulation of Eq. (A.29) leads to the same quadratic equation for the ratios r_{11}/r_{21} and r_{12}/r_{22}:

$$t_{21}(r_{11}/r_{21})^2 + (t_{22} - t_{11})(r_{11}/r_{21}) - t_{12} = 0,$$

(A.30)

$$t_{21}(r_{12}/r_{22})^2 + (t_{22} - t_{11})(r_{12}/r_{22}) - t_{12} = 0.$$

(A.31)

It can be shown that one solution of the quadratic equation is (r_{11}/r_{21}) and the second one is (r_{12}/r_{22}). As the third parameter we choose

$$e^{2jkl} = \frac{t_{21}(r_{12}/r_{22}) + t_{22}}{t_{12}(r_{11}/r_{21})^{-1} + t_{11}},$$

(A.32)

where the two solutions of the quadratic equations appear. From Eq. (A.32), the root choice for (r_{11}/r_{21}) and (r_{12}/r_{22}) can be made from the condition $\left| e^{2jkl} \right| > 1$, as the line is lossy. A more practical criterion for the root choice (for a network A with $\mathbf{R}_a$ that has higher transmission than reflection) turns out to be $|r_{12}/r_{22}| < |r_{11}/r_{21}|$.

To continue with the evaluation of the reflect data, the matrices $\mathbf{R}_a$, $\mathbf{R}_b$ and $\mathbf{R}_t$ are normalized with respect to their x_{22}-matrix elements according to

$$\mathbf{R}_a = r_{22} \begin{pmatrix} a & b \\ c & 1 \end{pmatrix}, \quad \mathbf{R}_b = \rho_{22} \begin{pmatrix} \alpha & \beta \\ \gamma & 1 \end{pmatrix}, \quad \mathbf{R}_t = g \begin{pmatrix} d & e \\ f & 1 \end{pmatrix}.$$

(A.33)

The measured or simulated reflection of a load Γ transformed through the network A is

$$w_1 = \frac{a\Gamma + b}{c\Gamma + 1}.\tag{A.34}$$

The already determined parameters are $b = (r_{12}/r_{22})$ and $a/c = (r_{11}/r_{21})$. From Eq. (A.34) the parameter a can be obtained by

$$a = \frac{w_1 - b}{\Gamma(1 - w_1 c/a)}.\tag{A.35}$$

If the reflection of the load Γ would be known, then a is determined and with it R_a except for the normalization parameter r_{11}. However, in the TRL procedure, Γ is by hypothesis unknown. Thus, we need to continue to determine the coefficients of network B.

Multiplying Eq. (A.26) from the left with $\mathbf{R}_a^{-1}$ and expanding the matrix equation leads to the following parameter relations for network B:

$$\gamma = \frac{f - dc/a}{1 - ec/a}, \quad \beta/a = \frac{e - b}{d - bf} \quad a\alpha = \frac{d - bf}{1 - ec/a},\tag{A.36}$$

where the parameters on the right hand sides b, a/c have been determined and d, e, f are known from the measured or simulated $\mathbf{R}_t$. Similarly to Eq. (A.35) for network A, evaluating the reflection after network B leads to

$$\alpha = \frac{w_2 + \gamma}{\Gamma(1 + w_2\beta/\alpha)}.\tag{A.37}$$

The unknown load reflection Γ can be eliminated using Eq. (A.35) and combined with Eq. (A.36) to yield:

$$a = \pm\sqrt{\frac{(w_1 - b)(1 + w_2\beta/\alpha)(d - bf)}{(w_2 + \gamma)(1 - w_1 c/a)(1 - ec/a)}}\tag{A.38}$$

and

$$\alpha = \frac{d - bf}{a(1 - ec/a)}.\tag{A.39}$$

The correct sign of a in Eq. (A.38) can be found by solving Eq. (A.35) for Γ and substituting a in the equation,

$$\Gamma = \frac{w_1 - b}{a(1 - w_1 c/a)},\tag{A.40}$$

and with the knowledge about the nominal value of Γ, the correct sign of a can be inferred.

Up to now, the parameters a, b, c and α, β, γ of the networks A and B have been determined, together with the line characteristics e^{2jkl} and the reflection characteristics Γ. However, the normalization parameters r_{22} and ρ_{22} are still unknown. It can be shown that only the product $r_{22}\rho_{22}$ is needed to achieve the calibration, and the product can be found by evaluating one of the equations of the matrix Eq. (A.26), e. g.

$$r_{22}\rho_{22} = \frac{g}{1 + c\beta}.\tag{A.41}$$

After the calibration coefficients have been determined, the calibration can be performed, where from the measured or simulated characteristics $\mathbf{R}_{\text{total}}$ including the networks A and B, see Fig. A.1(a),

$$\mathbf{R}_{\text{total}} = \begin{pmatrix} R_{11} & R_{12} \\ R_{21} & R_{22} \end{pmatrix} \tag{A.42}$$
$$= \mathbf{R}_{\text{a}}\mathbf{R}_{\text{Dut}}\mathbf{R}_{\text{b}},$$

the characteristics of the device under test can be obtained by:

$$\mathbf{R}_{\text{Dut}} = \mathbf{R}_{\text{a}}^{-1}\mathbf{R}_{\text{total}}\mathbf{R}_{\text{b}}^{-1} \tag{A.43}$$
$$= \frac{1}{r_{22}\rho_{22}\left(a-bc\right)\left(\alpha-\beta\gamma\right)} \times$$
$$\times \begin{pmatrix} R_{11}-bR_{21}-\gamma R_{12}+\gamma R_{22} & -\beta R_{11}+b\beta R_{21}+\alpha R_{12}-\alpha b R_{22} \\ -cR_{11}+aR_{21}+c\gamma R_{12}-a\gamma R_{22} & \beta cR_{11}-\beta aR_{21}-\alpha cR_{12}+\alpha aR_{22} \end{pmatrix}.$$

As a last step, the wave cascading matrix $\mathbf{R}_{\text{Dut}} = \begin{pmatrix} R_{11,\text{Dut}} & R_{12,\text{Dut}} \\ R_{21,\text{Dut}} & R_{22,\text{Dut}} \end{pmatrix}$ can be again transformed into the scattering matrix $\mathbf{S}_{\text{Dut}}$:

$$\mathbf{S}_{\text{Dut}} = \frac{1}{R_{22,\text{Dut}}}\begin{pmatrix} R_{12,\text{Dut}} & R_{11,\text{Dut}}R_{22,\text{Dut}}-R_{12,\text{Dut}}R_{21,\text{Dut}} \\ 1 & -R_{21,\text{Dut}} \end{pmatrix}. \tag{A.44}$$

A.5 Coupling of Counter-Propagating Modes

The model for coupling of two counter-propagating modes in a PC-WG is based on the *non-orthogonal coupled-mode theory* [101], [102]. Strictly speaking, the formalism is valid only for weak perturbations, which is not the case for the strongly coupled modes in WGs with high index contrast. However, the basic physical behavior is well explained with the model, and the coupling phenomena can also be well quantified if effective coupling parameters are used that fit measured or simulated characteristics.

In a single WG with different coupled modes or in two coupled WGs, the total $\mathbf{E}$ and $\mathbf{H}$-field distribution is represented by a trial solution, which is the superposition of the uncoupled vector modes $\hat{\mathbf{E}}_i$, $\hat{\mathbf{H}}_i$, $i = 1 \ldots N$:

$$\mathbf{E} = \sum_{i=1}^{N} a_i \hat{\mathbf{E}}_i, \tag{A.45}$$

$$\mathbf{H} = \sum_{i=1}^{N} a_i \hat{\mathbf{H}}_i. \tag{A.46}$$

The complex mode amplitudes a_i are dependent on the coordinate z, which is the propagation direction. The propagation of mode i with propagation constant β_i in positive z-direction is described by the propagation factor $\exp\left(\mathrm{j}\,\omega t - \mathrm{j}\,\beta_i z\right)$. The *coupled mode equations* determine the field evolution in the WG and can be obtained by substituting Eqs. (A.45)-(A.46) into Maxwell's Equations. The coupled mode equations can be written in the following matrix form:

$$\mathbf{P}\frac{\mathrm{d}}{\mathrm{d}z}\mathbf{A} = -\mathrm{j}\,\mathbf{H}\mathbf{A} \quad \text{or} \tag{A.47}$$

$$\frac{\mathrm{d}}{\mathrm{d}z}\mathbf{A} = -\mathrm{j}\,\mathbf{P}^{-1}\mathbf{H}\mathbf{A}. \tag{A.48}$$

This is a differential equation for the column vector of mode amplitudes $\mathbf{A} = (a_i)$ as a function of the z-position. The coefficients of the matrix $\mathbf{H} = (H_{ij})$, of the power matrix $\mathbf{P} = (P_{ij})$ and of the couling matrix $\kappa = (\kappa_{ij})$ are defined as

$$H_{ij} = P_{ij}\beta_j + \kappa_{ij}, \tag{A.49}$$

$$P_{ij} = \frac{1}{4}\int \left(\hat{\mathbf{E}}_i^* \times \hat{\mathbf{H}}_j + \hat{\mathbf{E}}_j \times \hat{\mathbf{H}}_i^*\right) \cdot \mathbf{e}_z \,\mathrm{d}A, \tag{A.50}$$

$$\kappa_{ij} = \frac{1}{4}\omega\varepsilon_0 \int \left(n^2 - n_j^2\right)\hat{\mathbf{E}}_i^* \cdot \hat{\mathbf{E}}_j \,\mathrm{d}A. \tag{A.51}$$

The propagation constants β_j may be positive or negative depending on the phase velocity direction. The energy velocity direction corresponds to the group velocity direction and is represented by the sign of P_{jj} in the power matrix. Note that the power matrix elements P_{ij} look very similar to the orthogonality relation, Eqs. (1.17) and (A.15), but that the fields denoted by i and j need not be orthogonal here. In the equation for the coupling elements κ_{ij}, $n = n(x,y)$ is the refractive index of the system with coupling, whereas $n_j = n_j(x,y)$ is the refractive index of the jth WG in the uncoupled system. The power matrix $\mathbf{P}$ is Hermitian with its elements satisfying $P_{ij} = P_{ji}^*$, and for a lossless system also $\mathbf{H}$ needs to be Hermitian with $H_{ij} = H_{ji}^*$. This implies that the coupling matrix κ is in general not Hermitian

In the coupled system there exist field distributions that propagate along z-direction without spatial variation of the field profile. These so-called supermodes are a superposition of the originally uncoupled modes. The supermodes fulfill the equation

$$\mathbf{A}(z) = \mathbf{A}_0 \, \mathrm{e}^{-j\gamma z}, \tag{A.52}$$

where γ is the (possibly complex) propagation constant of the supermode, and $\mathbf{A}_0$ are the expansion coefficients of the supermode in the basis of the unperturbed modes. If the ansatz Eq. (A.52) is substituted in the coupled mode equations Eq. (A.47), an eigenvalue equation follows:

$$\left(\mathbf{P}^{-1}\mathbf{H}\right)\mathbf{A}_0 = \gamma \mathbf{A}_0. \tag{A.53}$$

The eigenvectors and eigenvalues of this equation are the supermode coefficients $\mathbf{A}_0$ and the propagation constants γ. The number of eigenmodes and eigenvalues is given by the rank of the matrix $\left(\mathbf{P}^{-1}\mathbf{H}\right)$, which corresponds to the number N of originally uncoupled modes.

As a special case, we now consider two PC-WG modes $\hat{\mathbf{E}}_{1,2}$, $\hat{\mathbf{H}}_{1,2}$ which have opposite group velocity directions and will be coupled, where the coupling should not vary along z. Note that although the group velocities of the two modes point in different directions, the propagation constants need not have different sign; both modes can have positive propagation constants in the first Brillouin zone. We assume the modes to be orthogonal, $P_{12} = P_{21} = 0$, and normalize the mode powers such that mode 1 with positive group velocity gives $P_{11} = 1$, while mode 2 with negative group velocity gives $P_{22} = -1$, which results in $\mathbf{P} = \begin{pmatrix} 1 & 0 \\ 0 & -1 \end{pmatrix}$. The matrix $\left(\mathbf{P}^{-1}\mathbf{H}\right)$ for that case is

$$\mathbf{P}^{-1}\mathbf{H} = \begin{pmatrix} 1 & 0 \\ 0 & -1 \end{pmatrix}\begin{pmatrix} \beta_1 + \kappa_{11} & \kappa_{12} \\ \kappa_{21} & -\beta_2 + \kappa_{22} \end{pmatrix} = \begin{pmatrix} \beta_1 + \kappa_{11} & \kappa_{12} \\ -\kappa_{21} & \beta_2 - \kappa_{22} \end{pmatrix}. \tag{A.54}$$

The Hermitian behavior of $\mathbf{H}$ requires $\kappa_{12} = \kappa_{21}^*$, and $\beta_1 + \kappa_{11}$ and $-\beta_2 + \kappa_{22}$ to be real. We now denote $\beta_1 + \kappa_{11} = \beta_1'$ and $\beta_2 - \kappa_{22} = \beta_2'$. The two eigenvalues $\gamma_{a,b}$ of $\mathbf{P}^{-1}\mathbf{H}$ can be calculated as

$$\gamma_{a,b} = \frac{\beta_1' + \beta_2'}{2} \pm \frac{1}{2}\sqrt{\left(\beta_1' - \beta_2'\right)^2 - 4|\kappa_{12}|^2}. \tag{A.55}$$

The values $\gamma_{a,b}$ are the propagation constants of the modes in the coupled system, and they can be calculated from the propagation constants $\beta_{1,2}$ of the modes without coupling and the coupling constants κ_{11}, κ_{12} and κ_{22}. As the values β_1' and β_2' are real (assuming a lossless system), the values of $\gamma_{a,b}$ will become complex if the radical is negative, i. e. for $|\beta_1' - \beta_2'| < 2|\kappa_{12}|$. In that case, the real and and imaginary parts are $\gamma_{a,b} = \frac{1}{2}(\beta_1' + \beta_2') \pm \frac{1}{2}j\sqrt{\left|(\beta_1' - \beta_2')^2 - 4|\kappa_{12}|^2\right|}$, which has the consequence that the field amplitudes increase or decrease exponentially with position z according to $\left|\exp\left(-j\gamma_{a,b}z\right)\right| = \exp\left(\pm\frac{1}{2}\sqrt{\left|(\beta_1' - \beta_2')^2 - 4|\kappa_{12}|^2\right|}z\right)$. This means that no truly guided modes exist in the corresponding frequency range, and such a behavior is called a mode gap or a mini-stop band in a PC-WG. The reason for interaction between the counter-propagating modes can be either an intrinsic coupling mechanism like structure asymmetry or parallel-coupling, or it can be an extrinsic mechanism like structural disorder, see Sections 1.4

and 2.2. As an example, we will show the case of two parallel-coupled PC-WGs, where the uncoupled modes of the individual PC-WGs are counter-propagating (see also Subsection 2.1.4); the band diagram is shown in Fig. A.2. The supermodes of the coupled WGs exhibit a stop band in the region where the difference between the propagation constants of the two uncoupled modes is small. The coupling coefficients can be approximately determined by applying a parameter fit of the simulated band diagrams to the model.

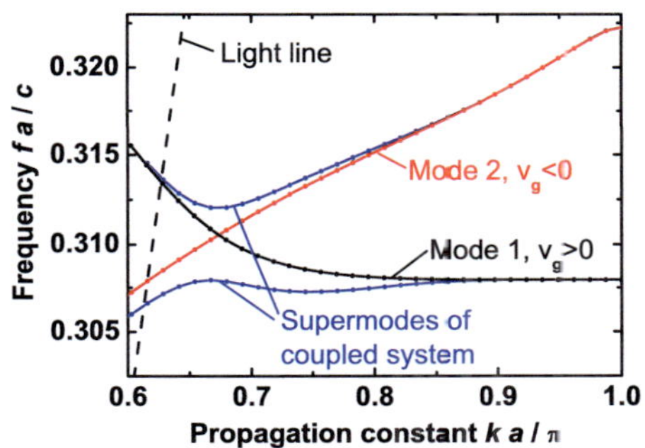

Fig. A.2. Parallel-coupling of two PC-WGs with counter-propagating modes. Displayed are the bands of the two uncoupled WGs exhibiting opposite group velocity directions, and the bands of the supermodes of the coupled system. In the coupled case, a stop gap arises, which is the frequency range without modes.

To study the behavior of the coupled WG if power is launched into one of the WGs inside the mode gap, we also need to take a closer look at the eigenvectors of the coupled system. The eigenvectors that correspond to the eigenvalues $\gamma_{a,b}$ are denoted as $\mathbf{A}_{0,a} = \begin{pmatrix} a_{1,a} \\ a_{2,a} \end{pmatrix}$ and $\mathbf{A}_{0,b} = \begin{pmatrix} a_{1,b} \\ a_{2,b} \end{pmatrix}$. The fields of the two supermodes can be written as the superposition of the formerly uncoupled modes

$$\mathbf{E}_a(z) = \left(a_{1,a}\hat{\mathbf{E}}_1 + a_{2,a}\hat{\mathbf{E}}_2\right)\exp\left(-\gamma_a z\right), \tag{A.56}$$

$$\mathbf{E}_b(z) = \left(a_{1,b}\hat{\mathbf{E}}_1 + a_{2,b}\hat{\mathbf{E}}_2\right)\exp\left(-\gamma_b z\right). \tag{A.57}$$

The same follows for the magnetic fields. The total electric field $\mathbf{E}$ in the coupled system can be represented in the basis of the supermodes, where the expansion coefficients are c_a and c_b

$$\mathbf{E}(z) = c_a\mathbf{E}_a(z) + c_b\mathbf{E}_b(z) \tag{A.58}$$

$$= \left[c_a a_{1,a}\exp\left(-\gamma_a z\right) + c_b a_{1,b}\exp\left(-\gamma_b z\right)\right]\hat{\mathbf{E}}_1 \tag{A.59}$$

$$+ \left[c_a a_{2,a}\exp\left(-\gamma_a z\right) + c_b a_{2,b}\exp\left(-\gamma_b z\right)\right]\hat{\mathbf{E}}_2.$$

The boundary conditions of the system determine c_a and c_b. We choose to inject power into mode 1 at position $z = 0$ of the two infinitely long coupled WGs. At the end of the WG $z =$

∞ no fields can exist, which requires the exponentially increasing function in Eq. (A.59) to vanish, which is only satisfied by choosing $c_b = 0$. The field inside the WG is then given by $\mathbf{E}(z) = c_a a_{1,a} \exp(-\gamma_a z) \, \hat{\mathbf{E}}_1 + c_a a_{2,a} \exp(-\gamma_a z) \, \hat{\mathbf{E}}_2$, which is the sum of the modes of the two originally uncoupled WGs with an exponential decay. The power injected in the first WG mode is transferred into the counter-propagating mode of the other WG, which means that all injected power will be reflected from the coupled WG system.

A.6 Loss Modeling

Here, we sketch the solution of the differential equations that govern the power flow in a PC-WG with backscattering α_b and out-of-plane scattering α_o (or the sum of out-of-plane and in-plane scattering) as introduced in Subsection 2.2.1. Based on the coupled power equations [104], the differential equations to be solved are:

$$\frac{dP_r(z)}{dz} = -(\alpha_b + \alpha_o)P_r + \alpha_b P_l, \tag{A.60}$$

$$\frac{dP_l(z)}{dz} = (\alpha_b + \alpha_o)P_l - \alpha_b P_r, \tag{A.61}$$

where the total loss is $\alpha_t = \alpha_b + \alpha_o$. The boundary conditions that have to be met are that the injected power at the beginning of the WG (position $z = 0$) is normalized to 1, $P_r(0) = 1$, and the reflected power at the end of the WG (position $z = L$) is zero, $P_l(L) = 0$. The Eqs. (A.60)-(A.61) can be solved by using the Laplace transform (s-transform), where a derivative is converted into a multiplication with s and an additional summand that takes the boundary conditions at position 0, $P_r(0)$ and $P_l(0)$, into account:

$$s\, p_r(s) - P_r(0) = -\alpha_t\, p_r(s) + \alpha_b\, p_l(s) \tag{A.62}$$

$$s\, p_l(s) - P_l(0) = \alpha_t\, p_l(s) - \alpha_b\, p_r(s) \tag{A.63}$$

Equation (A.63) can be reshaped to $p_l(s) = \frac{P_l(0) - \alpha_b\, p_r(s)}{s - \alpha_t}$ and substituted in Eq. (A.62) to isolate $p_r(s)$:

$$p_r(s) = \frac{\alpha_b P_l(0) - \alpha_t P_r(0)}{s^2 - (\alpha_t^2 - \alpha_b^2)} + \frac{s P_r(0)}{s^2 - (\alpha_t^2 - \alpha_b^2)}. \tag{A.64}$$

The result is already separated into fractions that can be transformed back by using the correspondences $\sinh(\alpha z) \circ\!\!-\!\!\bullet \frac{1}{s^2 - \alpha^2}$, $\cosh(\alpha z) \circ\!\!-\!\!\bullet \frac{s}{s^2 - \alpha^2}$, $\alpha \in \mathbb{R}$ and the following result is obtained:

$$P_r(z) = \frac{\alpha_b P_l(0) - \alpha_t P_r(0)}{\sqrt{\alpha_t^2 - \alpha_b^2}} \sinh\left(\sqrt{\alpha_t^2 - \alpha_b^2}\, z\right) + P_r(0)\cosh\left(\sqrt{\alpha_t^2 - \alpha_b^2}\, z\right). \tag{A.65}$$

Substituting $p_r(s)$ from Eq. (A.64) into $p_l(s) = \frac{P_l(0) - \alpha_b\, p_r(s)}{s - \alpha_t}$ leads to the solution for $p_l(s)$:

$$p_l(s) = \frac{\alpha_t P_l(0) - \alpha_b P_r(0)}{s^2 - (\alpha_t^2 - \alpha_b^2)} + \frac{s P_l(0)}{s^2 - (\alpha_t^2 - \alpha_b^2)}. \tag{A.66}$$

Using the same correspondences as before, the result is

$$P_l(z) = \frac{\alpha_t P_l(0) - \alpha_b P_r(0)}{\sqrt{\alpha_t^2 - \alpha_b^2}} \sinh\left(\sqrt{\alpha_t^2 - \alpha_b^2}\, z\right) + P_l(0)\cosh\left(\sqrt{\alpha_t^2 - \alpha_b^2}\, z\right). \tag{A.67}$$

From this equation, the conditions $P_l(L) = 0$ and $P_r(0) = 1$ allow solving for the backreflection $P_l(0)$:

$$P_l(0) = \frac{\alpha_b \tanh\left(\sqrt{\alpha_t^2 - \alpha_b^2}\, L\right)}{\alpha_t \tanh\left(\sqrt{\alpha_t^2 - \alpha_b^2}\, L\right) + \sqrt{\alpha_t^2 - \alpha_b^2}}. \tag{A.68}$$

If the substitution $\alpha \equiv \sqrt{\alpha_t^2 - \alpha_b^2}$ is made, Eq. (A.68) results in Eq. (2.3). The transmitted power $P_r(L)$ is obtained from Eq. (A.65), where $P_l(0)$ is known from Eq. (A.68):

$$P_r(L) = \frac{\alpha_b}{\sqrt{\alpha_t^2 - \alpha_b^2}} \sinh\left(\sqrt{\alpha_t^2 - \alpha_b^2}\, L\right) P_l(0)$$
$$+ \left[\cosh\left(\sqrt{\alpha_t^2 - \alpha_b^2}\, L\right) - \frac{\alpha_t}{\sqrt{\alpha_t^2 - \alpha_b^2}} \sinh\left(\sqrt{\alpha_t^2 - \alpha_b^2}\, L\right)\right]. \tag{A.69}$$

This again results in Eq. (2.4) with the substitution for α.

We also derive the solution of the simplified loss model:

$$\frac{\mathrm{d}P_r(z)}{\mathrm{d}z} = -(\alpha_b + \alpha_o)P_r, \tag{A.70}$$

$$\frac{\mathrm{d}P_l(z)}{\mathrm{d}z} = -\alpha_b P_r. \tag{A.71}$$

The first equation, Eq. (A.70) can be directly integrated to yield

$$P_r(z) = P_r(0)\,\mathrm{e}^{-\alpha_t L}. \tag{A.72}$$

This can be substituted into Eq. (A.70) and the solution can be found by integration:

$$\frac{\mathrm{d}P_l(z)}{\mathrm{d}z} = -\alpha_b P_r(0)\,\mathrm{e}^{-\alpha_t L}$$
$$\Rightarrow P_l(z) - P_l(0) = \frac{\alpha_b}{\alpha_t} P_r(0)\left(\mathrm{e}^{-\alpha_t z} - 1\right) \tag{A.73}$$

The boundary conditions $P_l(L) = 0$ and $P_r(0) = 1$ lead to the result for transmission and reflection, Eqs. (2.8)-(2.9):

$$P_l(0) = \frac{\alpha_b}{\alpha_t}\left(1 - \mathrm{e}^{-\alpha_t L}\right), \tag{A.74}$$

$$P_r(L) = \mathrm{e}^{-\alpha_t L}. \tag{A.75}$$

A.7 Determination of Modulator Characteristics

A.7.1 Propagation Equation and Field Interaction Factor Γ

For the derivation of the propagation equation for the optical mode in the modulator and the field interaction factor Γ we follow the formalism in [142]. The influence of the microwave field on the optical mode through the electro-optic effect is treated as a perturbation $\Delta\varepsilon$ on the permittivity of the material in Maxwell's equations:

$$\nabla \times \mathbf{E} = -\mu\, \partial\mathbf{H}/\partial t, \tag{A.76}$$

$$\nabla \times \mathbf{H} = \partial[(\varepsilon + \Delta\varepsilon)\mathbf{E}]/\partial t. \tag{A.77}$$

For a mode propagating inside a PC-WG along the z-direction, a complex envelope function A is defined,

$$\mathbf{E}(x,y,z,t) = A(z,t)\,\hat{\mathbf{E}}(x,y,z)\,e^{j(\omega_0 t - \beta_0 z)}, \tag{A.78}$$

$$\mathbf{H}(x,y,z,t) = A(z,t)\,\hat{\mathbf{H}}(x,y,z)\,e^{j(\omega_0 t - \beta_0 z)}. \tag{A.79}$$

The field distribution of the mode is taken at the optical carrier frequency $f_0 = \omega_0/(2\pi)$ and is assumed to be constant in the optical signal bandwidth centered at f_0. For the propagation constant β of the mode we assume $\beta(\omega) = \beta_0 + (\omega - \omega_0)v_{\mathrm{g,opt}}^{-1}$ neglecting chromatic dispersion. If Eqs. (A.78), (A.79) are substituted in Eqs. (A.76), (A.77), and both the mode orthogonality condition Eq. (1.17) and the slowly varying envelope approximation are used, a nonlinear propagation equation is obtained,

$$\frac{\partial A}{\partial z} + \frac{1}{v_{\mathrm{g,opt}}}\frac{\partial A}{\partial t} = j\,\omega_0\Gamma K U A, \qquad A = |A|\,e^{j(\Phi + \Delta\Phi)}. \tag{A.80}$$

In Eq. (A.80), the relations $A = A(z,t)$, $U = U(z,t)$, $\Phi = \Phi(z,t)$ and $\Delta\Phi = \Delta\Phi(z,t)$ hold. The field interaction factor Γ is defined to be

$$\Gamma = \int_{\mathrm{gap}} \frac{n_{\mathrm{EO}}}{Z_0}|\hat{E}_x|^2\,\mathrm{d}V \Big/ \int a\,\Re(\hat{\mathbf{E}} \times \hat{\mathbf{H}}^*)\cdot\mathbf{e}_z\,\mathrm{d}A \tag{A.81}$$

as already stated in Eq. (3.17), $K = \frac{\varepsilon_0}{2}r_{33}n_{\mathrm{EO}}^3 Z_0/W_{\mathrm{gap}}$ is a constant, and $U = U(z,t)$ is the voltage across the gap. It is generated by the microwave field that causes an electro-optic refractive index change $\Delta n = -r_{33}n_{\mathrm{EO}}^3 U/W_{\mathrm{gap}}$ and a resulting phase change $\Delta\Phi$.

A.7.2 Modulator Walk-Off Bandwidth

If we assume a sinusoidal temporal and spatial variation of U along the WG according to $U(z,t) = \hat{U}\cos\left(\omega_{\mathrm{el}}\left(t - z/v_{\mathrm{p,el}}\right)\right)$ with $v_{\mathrm{p,el}} \approx v_{\mathrm{g,el}}$, Eq. (A.80) can be solved, and the phase shift $\Delta\Phi$ of the optical signal after propagating along a WG section of length z is determined as:

$$\Delta\Phi(t) = \Delta\hat{\Phi}\cos\left(\omega_{\mathrm{el}}t + \varphi\right), \tag{A.82}$$

$$\varphi = -\omega_{\mathrm{el}}z/v_{\mathrm{g,opt}} + \Omega,$$

$$\Omega = \omega_{\mathrm{el}}\left(v_{\mathrm{g,opt}}^{-1} - v_{\mathrm{p,el}}^{-1}\right)L/2,$$

$$\Delta\hat{\Phi} = \omega_0\Gamma K L \hat{U}\,\frac{\sin\Omega}{\Omega}.$$

In the case of a DC electrical voltage where $\omega_{el} = 0$, $U = \hat{U}$, and $\sin\Omega/\Omega = 1$, this leads to a phase shift as also stated in Eq. (3.16):

$$\Delta\Phi = \Delta\hat{\Phi} = \omega_0\Gamma KL\hat{U} \qquad \text{for } \omega_{el} = 0 \tag{A.83}$$
$$= -\Gamma\Delta n k_0 L.$$

For amplitude modulation at $\omega_{el} \neq 0$, two phase modulators are operated in a MZ push-pull configuration. The optical signals in both arms experience phase shifts of $\Delta\Phi$ and $-\Delta\Phi$, respectively. The arms are biased at $\Delta\Phi_0 = \pi/4$, and an additional phase shift $\Delta\hat{\Phi} = \pi/4$ is needed to achieve complete extinction. The optical signal after the MZI is detected with a quadratic detector and generates an electrical current i:

$$i(t) \propto \cos^2\left(\Delta\Phi_0 + \Delta\Phi(t)\right) = \cos^2\left(\Delta\Phi_0 + \Delta\hat{\Phi}\cos\left(\omega_{el}t + \varphi\right)\right). \tag{A.84}$$

If the current spectrum $I(f)$ is evaluated at the frequency $f_{el} = \omega_{el}/(2\pi)$, the dependence on $\Delta\hat{\Phi}$ can be derived as $I(f_{el}) \propto J_1\left(2\Delta\hat{\Phi}(f_{el})\right)$, where J_1 is the Bessel-function of 1st order. The 3 dB modulation bandwidth Eq. (3.13) is the frequency f_{el} at which the magnitude of $I(f_{el})$ has decreased by a factor of $1/\sqrt{2}$ with respect to the DC case with $\Delta\hat{\Phi}(f_{el} = 0) = \pi/4$:

$$J_1\left(2\Delta\hat{\Phi}(f_{3dB})\right) = J_1\left(2\Delta\hat{\Phi}(f_{el} = 0)\right)/\sqrt{2}, \tag{A.85}$$
$$\hat{\Phi}(f_{3dB})/\hat{\Phi}(f_{el} = 0) \approx 0.564.$$

According to Eq. (A.82), we find $\hat{\Phi}(f_{3dB})/\hat{\Phi}(f_{el} = 0) = \sin\left(\Omega(f_{3dB})\right)/\Omega(f_{3dB})$, which leads to

$$\Omega(f_{3dB})/\pi = f_{el}\left(v_{g,\,opt}^{-1} - v_{p,\,el}^{-1}\right)L \approx 0.556, \tag{A.86}$$

thus proving Eq. (3.13) with the accurate factor of 0.556. With an analogous calculation, the walk-off bandwidth for counter-propagating electrical and optical signals can be determined, Eq. (3.14).

A.7.3 Modulator Bandwidth Limitations by *RC*-Effects

We estimate the modulator bandwidth limitations that are caused by electrical *RC*-effects. We discuss two different mechanisms that reduce the electrical voltage drop across the gap and lead to an associated 3 dB bandwidth. These are the generator-determined bandwidth and the parallel-loss determined bandwidth. We find that these bandwidth limitations do not play a significant role for the presented structures.

For the following numerical examples, the data of our specific MZM design for a 78 GHz modulation bandwidth and a drive voltage of 1 V were assumed, Fig. 3.12 and Fig. 3.13: Optical strip waveguide height $h = 220$ nm, PM section length and gap width $L = 80\,\mu$m and $W_{gap} = 150$ nm, linear refractive index of the PM section $n_{EO} = 1.6$ (assumed to be the same at optical and electrical frequencies), width of the doped silicon slabs $w = 3\,\mu$m, vacuum dielectric constant ε_0, vacuum speed of light c.

1) Generator-Determined Bandwidth

The CPWG with wave impedance $Z_L = 50\,\Omega$ be terminated with an open, Fig. A.3(a). We assume the PM section to be short compared to the modulation signal wavelength, so that

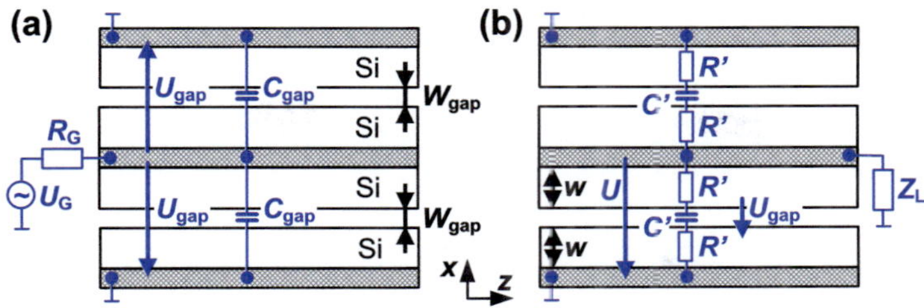

Fig. A.3. Electrical *RC*-effects. (a) Generator-determined limitation (b) Parallel-loss determined limitation. In (a), the electrically short PM section is represented by a lumped gap capacitance C_{gap}, and the voltage across the gap U_{gap} is only a fraction of the generator voltage U_G because of the generator impedance R_G. In (b), a voltage wave with constant amplitude $|U|$ travels in z-direction. The voltage U_{gap} across the gap (capacitance per length C') is reduced because of the finite resistivity of the doped silicon sections of width w (conductance per length (R'^{-1})).

the CPWG may be represented by a lumped gap capacitance $C_{gap} = \varepsilon_0 n^2_{poly} hL/W_{gap}$ for each PM. The voltage amplitude U_{gap} across the non-conductive gap W_{gap} (i.e., across the lumped capacitor $2C_{gap}$) decreases with modulation frequency f because of the generator impedance $R_G = 50\,\Omega$,

$$\frac{U_{gap}}{U_G} = \frac{1}{1+j\,\omega R_G 2 C_{gap}}, \qquad f^{(R_G C_{gap})}_{3\,dB} = \frac{1}{4\pi R_G C_{gap}}. \qquad (A.87)$$

At the limiting frequency $f^{(R_G C_{gap})}_{3\,dB}$ the gap voltage $|U_{gap}|$ drops to $|U_G|/\sqrt{2}$. Assuming the previously assigned data and neglecting electric fringing fields, the total gap capacitance amounts to $2C_{gap} = 5\,fF$, and $f^{(R_G C_{gap})}_{3\,dB} \approx 640\,GHz$ results. In view of the envisaged MZM bandwidth of $78\,GHz$, the generator-determined limitation is unimportant.

2) Parallel-Loss Determined Bandwidth

The CPWG be terminated with its wave impedance Z_L, Fig. A.3(b). Then, a modulating electrical wave traveling in z-direction has a spatially constant amplitude $|U|$. The resulting voltage amplitude $|U_{gap}|$ across the non-conductive gap W_{gap} (PM capacitance per length $C' = \varepsilon_0 n^2_{EO} h/W_{gap}$) is reduced because of the finite resistivity of the doped silicon. For silicon sections having a width w and a filling factor F taking the reduction of the effective conductance by the air holes into account (conductance per length $(R'^{-1} = \sigma F h/w)$), we obtain

$$\frac{U_{gap}}{U} = \frac{1}{1+j\,\omega 2 R' C'}, \qquad f^{(R'C')}_{3\,dB} = \frac{1}{4\pi R' C'} \qquad (A.88)$$

The limiting frequency $f^{(R'C')}_{3\,dB}$ denotes where the gap voltage $|U_{gap}|$ drops to $|U|/\sqrt{2}$. Assuming a conductivity $\sigma = 10\,\Omega^{-1}\,cm^{-1}$ of the doped ($n_D \approx 2 \times 10^{16}\,cm^{-3}$) silicon material and a filling factor $F = 0.67$, a bandwidth of $f^{(R'C')}_{3\,dB} \approx 120\,GHz$ results. In view of the envisaged MZM bandwidth of $78\,GHz$, the parallel-loss determined limitation does not play a significant role either.

3) Other Effects

Basically, one might think of a bandwidth limitation by the finite carrier transit time in the doped silicon slabs, see Fig. 3.13. This carrier drift time is determined by the slower carriers, which in silicon are holes, the maximum saturation velocity of which is about $0.6 \times 10^7 \, \text{cm}/\text{s}$. Over the $3 \, \mu\text{m}$ witdh of the silicon slab, this would lead to a response time in the order of $\tau_D = 50 \, \text{ps}$, resulting in a bandwidth in the order of $3 \, \text{GHz}$ only.

However, there is also the dielectric dielectric relaxation time τ_R, inside which any charge perturbation in the doped silicon slabs (induced by the modulating field) is screened by a shift of the whole carrier ensemble. The dielectric relaxation time $\tau_R = \varepsilon_0 \varepsilon_r / \sigma$ depends on the material's conductivity σ and permittivity $\varepsilon_0 \varepsilon_r$. For $\sigma = 10 \, \Omega^{-1} \text{cm}^{-1}$ as assumed above and with $\varepsilon_r = 12$, a value of $\tau_R = 0.1 \, \text{ps}$ is obtained.

For the resulting response time $\tau_{\text{res}}^{-1} = \tau_D^{-1} + \tau_R^{-1} \approx \tau_R^{-1}$, the *faster* of the two effects is relevant, and the bandwidth limited by the dielectric relaxation time would be $1.6 \, \text{THz}$. As a consequence, carrier transit times do not limit the modulator's performance.

A.8 Microwave Measurement Techniques

A.8.1 Continuous Wave Measurements

To obtain transmission, reflection, group velocity and dispersion characteristics of a device, continuous wave (CW) measurements are performed. The measurement instrument is an electrical vector network analyzer (VNA), which directly provides the scattering parameters (S-parameters) of a coaxial two-port device. During the measurement the frequency is swept, and the S-parameters are obtained for a number of frequency points within a specified range.

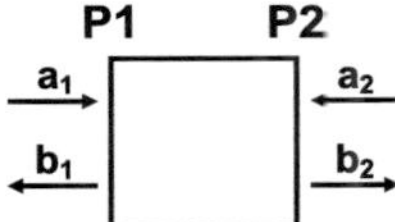

Fig. A.4. Notation for incoming $a_{1,2}$ and outgoing $b_{1,2}$ wave amplitudes of a two-port device with ports P1 and P2.

Figure A.4 depicts a linear device with two waveguide ports P1 and P2. The incoming and outgoing wave amplitudes are represented by complex numbers a_1, a_2 and b_1, b_2, respectively, where the subscript refers to the port number. The S-parameters link the incoming and the outgoing wave amplitudes of the device by the following matrix equation:

$$\begin{pmatrix} b_1 \\ b_2 \end{pmatrix} = \begin{pmatrix} S_{11} & S_{12} \\ S_{21} & S_{22} \end{pmatrix} \begin{pmatrix} a_1 \\ a_2 \end{pmatrix} \tag{A.89}$$

If the device is excited at port 1, $|S_{21}|^2$ is the fraction of power that is transmitted to port 2 and $|S_{11}|^2$ the fraction of power that is reflected at port 1. From the phase information $\varphi = \arg(S_{21})$ of the transmitted wave as a function of frequency, the group velocity can be calculated by $v_g = -2\pi L \left(d\varphi/df \right)^{-1}$, and the chromatic dispersion by $C = -f^2/\left(8\pi^3 cL\right)\left(d^2\varphi/df^2\right)$, where L is the device length.

In all measurements, the influences of the coaxial cables and the antennas are included in the results. In order to eliminate these influences, a through-reflect-line (TRL) calibration procedure was used, more details are given in Appendix A.4. The calibration defines the reference port planes of the coaxial-strip WG transitions (as also depicted in Fig. 4.14). Three calibration standards need to be measured in addition to the devices under test: A through standard, realized by a back-to-back butt coupling of two coaxial-strip WG transitions at the reference port planes, a reflection standard, in our case realized by a copper plate in the reference port plane, and a line standard, which is the connection of the two coaxial-strip WG transitions with a section of a matched dielectric strip-WG in-between. Photographs of the three calibration measurement setups are shown in Fig. A.5. Note that in contrast to the microwave measurements, the reference port planes for the disorder simulations in Section 2.2 are located inside the PC sections, see Fig. 4.26.

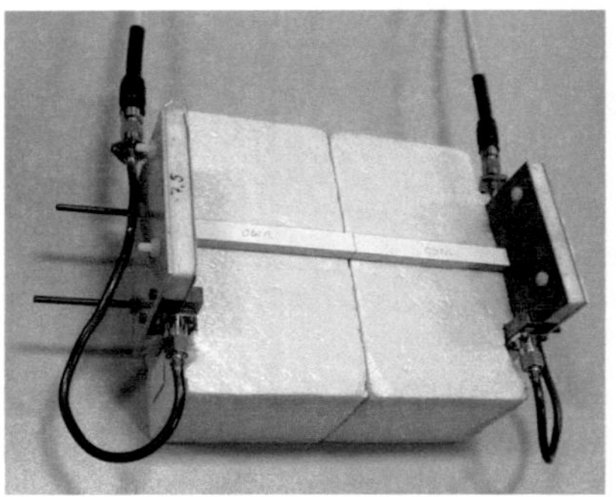

(a) Through standard.

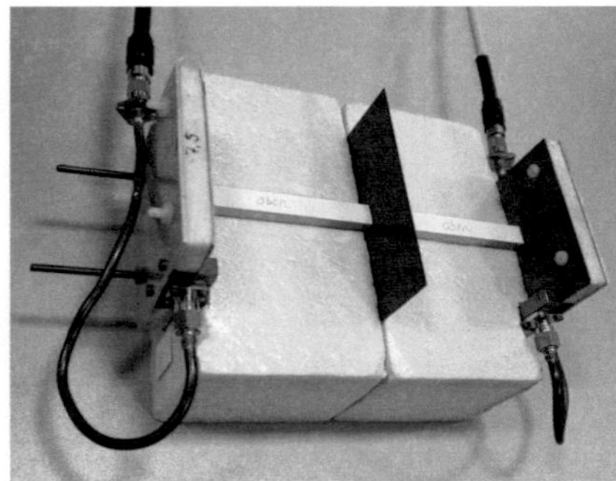

(b) Reflection standard.

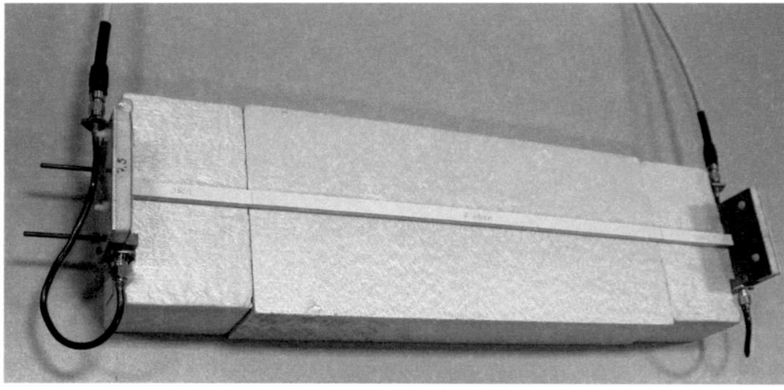

(c) Line standard.

Fig. A.5. Measurement setup for the three standards needed for TRL calibration. (a) The through standard is realized by the back-to-back coupling of the two coaxial-strip WG transitions at the reference port planes. (b) The reflection standard is realized by a copper plate at the reference port planes. (c) For the line standard, the two coaxial-strip WG transitions are connected with a section of a matched strip-WG in-between.

A.8.2 Pulse Measurements

To model the behavior of the devices under test embedded in a fast communication system, we conducted broadband pulse transmission experiments. The measurement setup with symmetrically arranged coaxial-strip WG transitions is displayed in Fig. A.6. A CW carrier generated with a microwave synthesizer is modulated by a Gaussian-shaped pulse, transmitted through the microwave model, and detected with a square law detector. The carrier frequency is varied in the region of interest around $f' = 10\,\mathrm{GHz}$. The pulse duration is about 10 ns, and if the time axis is scaled to the optical regime, this corresponds to $10\,\mathrm{ns}/20,000 = 0.5\,\mathrm{ps}$. In Fig. A.6, a microwave model of a PC-WG is depicted, fed by a strip-WG and excited with a slot antenna. In the optical domain, a pulse transmission setup would look very similar, but with optical components instead of microwave components.

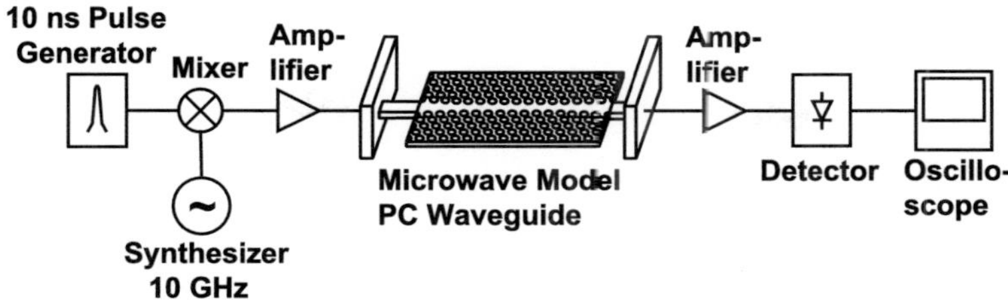

Fig. A.6. Measurement setup for pulse transmission in the microwave model. With a microwave mixer, a 10 ns pulse is modulated on the 10 GHz carrier and amplified before it is injected into the microwave model. At the output of the model, the signal is amplified again, before it is detected and visualized with an oscilloscope.

A.8.3 Near-Field Measurements

In the microwave model, it is easily possible to measure the field distribution in a plane above but close to the PC slab. We set up a measurement system with a moveable field probe that scans the surface in two dimensions with an equivalent optical resolution of 5 nm, see Fig. A.7. With differently arranged dipole or loop probes it is possible to measure the amplitude of all **E**- and **H**-field components. These measurements are the much refined microwave equivalent of scanning near field optical microscopy (SNOM) experiments in the optical domain [88].

Fig. A.7. Measurement setup for the 2D near-field distribution. A moveable field probe scans the surface in two dimensions, and various probes allow measuring the different field components.

Glossary

Acronyms

2D	Two-dimensional
3D	Three-dimensional
CMOS	Complementary metal oxide semiconductor
CPO	Coherent population oscillation
CPWG	Coplanar waveguide
CROW	Coupled-resonator optical waveguide
CW	Continuous wave
DCF	Dispersion compensating fiber
DUT	Device under test
EIT	Electromagnetically induced transparency
FCA	Free-carrier absorption
FDTD	Finite difference time domain
FIT	Finite integration technique
FWHM	Full width at half maximum
GME	Guided mode expansion
MZ	Mach-Zehnder
MZI	Mach-Zehnder interferometer
MZM	Mach-Zehnder modulator
PBG	Photonic bandgap
PC	Photonic crystal
PM	Phase modulator
PWE	Plane wave expansion
RMS	Root mean square
SBS	Stimulated Brillouin scattering
SEM	Scanning electron microscope
SMF	Single mode fiber
SOI	Silicon-on-insulator
SRS	Stimulated Raman scattering
TE	Transverse electric

TIR	Total internal reflection
TM	Transverse magnetic
TPA	Two-photon absorption
TRL	Thru-reflect-line
WG	Waveguide

Symbols

∇ Nabla operator

$\nabla\times$ Rotation operator, Eq. (1.1)

$\nabla\cdot$ Divergence operator, Eq. (1.3)

ΔX Small Variation of the physical value X with respect to its nominal value X_0, $|\Delta X/X_0| \ll 1$

$\Re$ Real part of a complex number

α Power attenuation coefficient or propagation loss factor, unit $1/\text{m}$, Section 2.2

α_b Backscattering loss factor, unit $1/\text{m}$, Fig. 1.11

α_t Total loss factor, unit $1/\text{m}$, Fig. 1.11

α_o Out-of-plane loss factor, unit $1/\text{m}$, Fig. 1.11

β Propagation constant of a mode, unit $1/\text{m}$, Eq. (1.14)

$\beta^{(n)}$ nth derivative of β with respect to ω, $\beta^{(n)} = \text{d}^n\beta/\text{d}\omega^n$, unit s^n/m, Eq. (1.21)

ε_0 Permittivity of vacuum, $\varepsilon_0 = 8.85419 \times 10^{-12}\,\text{As}/(\text{Vm})$, Eq. (1.2)

ε_r Relative permittivity or dielectric constant, Eq. (1.2)

Γ Field interaction factor, Eq. (3.17)

λ_0 Free-space wavelength, $\lambda_0 = c/f$, unit m, Section 1.3

μ_0 Permeability of vacuum, $\mu_0 = 4\pi \times 10^{-7}\,\text{Vs}/(\text{Am}) = {} = 1.25664 \times 10^{-6}\,\text{Vs}/(\text{Am})$, Eq. (1.1)

φ Optical phase

$\Delta\Phi$ Optical phase shift, Eq. (3.16)

σ Disorder level, root mean square of disorder, unit m

$\chi^{(2)}_{ijk}$ Second order nonlinear susceptibility, unit $(\text{m}/\text{V})^2$, Eq. (1.31)

ω Angular frequency, $\omega = 2\pi f$, unit $1/\text{s}$, Eq. (1.7)

a Lattice constant of a photonic crystal, unit m, Section 1.2

$\mathbf{B}$ Magnetic flux density vector, unit Vs/m^2, Eq. (1.4)

C Chromatic dispersion, unit $\text{ps}/(\text{mm}\,\text{nm})$, Eq. (1.24)

c Vacuum speed of light, $c = 1/\sqrt{\varepsilon_0\mu_0} = 2.99792 \times 10^8\,\text{m}/\text{s}$

$C_{\text{dly-bw}}$ Delay-bandwidth product of a delay line, Eq. (4.1)

$\mathbf{D}$ Electric displacement vector, unit As/m^2, Eq. (1.3)

$\mathbf{E}$ Electric field vector, unit V/m, Eq. (1.1)

$\hat{\mathbf{E}}$ Electric field profile of a mode, unit V/m, Eq. (1.14)

$\mathbf{e}_{x,y,z}$ Unit vector in basic coordinate direction x, y or z

f	Frequency, unit Hz
$f_{3\,\mathrm{dB}}$	3 dB modulation bandwidth, unit Hz, Eq. (3.12)
Δf	Bandwidth, unit Hz
$\mathbf{G}$	Reciprocal lattice vectors of a photonic crystal, unit $1/\mathrm{m}$, Section 1.2
$\mathbf{H}$	Magnetic field vector, unit A/m, Eq. (1.1)
$\hat{\mathbf{H}}$	Magnetic field profile of a mode, unit A/m, Eq. (1.15)
h	Slab height of a structure, unit m
j	Imaginary unit, $\mathrm{j}^2 = -1$
$\mathbf{k}$	Wave vector, unit $1/\mathrm{m}$, Eq. (1.19)
k	Propagation constant, unit $1/\mathrm{m}$, Eq. (1.20)
L, l	Length of a structure, unit m
n	Refractive index, $n = \varepsilon_r^2$
n_{eff}	Effective refractive index, Eq. 1.16
n_{g}	Group index, Section (1.3)
P	Power of a mode, unit W, Eq. (1.18)
$\mathbf{R}$	Lattice vectors of a photonic crystal, unit m, Section 1.2
$\mathbf{r}$	Spatial coordinate vector, $\mathbf{r} = (x, y, z)$, unit m
r	Radius of a photonic crystal hole, unit m
r_{ij}	Electro-optic coefficient, unit m/V, Eq. (1.34)
S_{ij}	Scattering parameters of a device, Eq. (A.24)
t	Time, unit s
t_{g}	Group delay, $t_{\mathrm{g}} = L/v_{\mathrm{g}}$, unit s, Section 1.3
U	Voltage, unit V
U_{π}	Required modulation voltage to generate a π phase shift, unit V, Eq. (3.12)
$\hat{U}$	Amplitude of modulation voltage, unit V, Section 3.3.4
$\mathbf{u}$	Periodic part of a photonic crystal mode, Eq. (1.19)
v_{p}	Phase velocity, unit m/s, Eq. (1.22)
v_{g}	Group velocity, unit m/s, Eq. (1.23)
W	Geometrical width, unit m
x, y, z	Spatial coordinates, unit m, Fig. 1.1
Z_0	Wave impedance of free space, $Z_0 = \sqrt{\mu_0/\varepsilon_0} = 376.73\,\Omega$, Eq. (3.17)

Bibliography

[1] J. M. Brosi, C. Koos, L. C. Andreani, M. Waldow, J. Leuthold, and W. Freude, "High-speed low-voltage electro-optic modulator with a polymer-infiltrated silicon photonic crystal waveguide," *Opt. Express*, vol. 16, pp. 4177–4191, 2008.

[2] J. M. Brosi, J. Leuthold, and W. Freude, "Microwave-frequency experiments validate optical simulation tools and demonstrate novel dispersion-tailored photonic crystal waveguides," *J. Lightw. Technol.*, vol. 25, pp. 2502–2510, 2007.

[3] E. Yablonovitch, "Inhibited spontaneous emission in solid-state physics and electronics," *Phys. Rev. Lett.*, vol. 58, pp. 2059–2062, 1987.

[4] S. John, "Strong localization of photons in certain disordered dielectric superlattices," *Phys. Rev. Lett.*, vol. 58, pp. 2486–2489, 1987.

[5] E. Yablonovitch and T. J. Gmitter, "Photonic band structure: The face-centered-cubic case," *Phys. Rev. Lett.*, vol. 63, pp. 1950–1953, 1989.

[6] E. Yablonovitch and T. J. Gmitter, "Donor and acceptor modes in photonic band structures," *Phys. Rev. Lett.*, vol. 67, pp. 3380–3383, 1991.

[7] J. D. Jackson, *Classical Electrodynamics*. New York: John Wiley and Sons, 1998.

[8] J. A. Kong, *Electromagnetic Wave Theory*. Cambridge. Massachusetts: EMW Publishing, 2000.

[9] J. D. Joannopoulos, R. D. Meade, and J. N. Winn, *Photonic Crystals: Molding the Flow of Light*. New Jersey: Princeton University Press, 1995.

[10] D. Marcuse, *Light Transmission Optics*. New York: Van Nostrand Reinhold, 1972.

[11] S. G. Johnson, S. Fan, P. R. Villeneuve, J. D. Joannopoulos, and L. A. Kolodziejski, "Guided modes in photonic crystal slabs," *Phys. Rev. E*, vol. 60, pp. 5751–5758, 1999.

[12] M. Loncar, D. Nedeljkovic, T. Doll, J. Vuckovic, A. Scherer, and T. P. Pearsall, "Waveguiding in planar photonic crystals," *Appl. Phys. Lett.*, vol. 77, pp. 1937–1939, 2000.

[13] S. G. Johnson, P. R. Villeneuve, S. Fan, and J. D. Joannopoulos, "Linear waveguides in photonic-crystal slabs," *Phys. Rev. B*, vol. 62, pp. 8212–8222, 2000.

[14] M. Notomi, K. Yamada, A. Shinya, J. Takahashi, C. Takahashi, and I. Yokohama, "Extremely large group-velocity dispersion of line-defect waveguides in photonic crystal slabs," *Phys. Rev. Lett.*, vol. 87, p. 253902, 2001.

[15] M. Soljacic and J. D. Joannopoulos, "Enhancement of nonlinear effects using photonic crystals," *Nature Materials*, vol. 3, pp. 211–219, 2004.

[16] R. S. Jacobsen, K. N. Andersen, P. I. Borel, J. Fage-Pedersen, L. H. Frandsen, O. Hansen, M. Kristensen, A. V. Lavrinenko, G. Moulin, H. Ou, C. Peucheret, B. Sigri, and A. Bjarklev, "Strained silicon as a new electro-optic material," *Nature*, vol. 441, pp. 199–202, 2006.

[17] T. Baehr-Jones, M. Hochberg, G. Wang, R. Lawson, Y. Liao, S. P. A., L. Dalton, A. K.-Y. Jen, and A. Scherer, "Optical modulation and detection in slotted silicon waveguides," *Opt. Express*, vol. 13, pp. 5216–5226, July 2005.

[18] R. G. DeCorby, N. Ponnampalam, M. M. Pai, H. T. Nguyen, P. K. Dwivedi, T. J. Clement, C. J. Haugen, J. N. McMullin, and S. O. Kasap, "High index contrast waveguides in chalcogenide glass and polymer," *IEEE J. Sel. Top. Quantum Electron.*, vol. 11, pp. 539–546, 2005.

[19] R. Claps, D. Dimitopoulos, V. Raghunathan, Y. Han, and B. Jalali, "Observation of stimulated Raman amplification in silicon waveguides," *Opt. Express*, vol. 11, pp. 1731–1739, 2003.

[20] H. Rong, S. Xu, Y. H. Kuo, V. Sih, O. Cohen, O. Raday, and M. Paniccia, "Low-threshold continuous-wave Raman silicon laser," *Nature Photonics*, vol. 1, pp. 232–237, 2007.

[21] M. A. Foster, A. C. Turner, J. E. Sharping, B. S. Schmidt, M. Lipson, and A. L. Gaeta, "Broad-band optical parametric gain on a silicon photonic chip," *Nature*, vol. 441, pp. 960–963, 2006.

[22] Y. Tanaka, Y. Sugimoto, N. Ikeda, H. Nakamura, K. Asakawa, K. Inoue, and S. G. Johnson, "Group velocity dependence of propagation losses in single-line-defect photonic crystal waveguides on GaAs membranes," *Electron. Lett.*, vol. 40, pp. 174–176, 2004.

[23] A. Mekis, S. Fan, and J. D. Joannopoulos, "Bound states in photonic crystal waveguides and waveguide bends," *Phys. Rev. B*, vol. 58, pp. 4809–4817, 1998.

[24] S. Olivier, M. Rattier, H. Benisty, C. Weisbuch, C. J. M. Smith, R. M. De La Rue, T. F. Krauss, U. Oesterle, and R. Houdr, "Mini-stopbands of one-dimensional system: The channel waveguide in a two-dimensional photonic crystal," *Phys. Rev. B*, vol. 63, p. 113311, 2001.

[25] M. Agio and C. M. Soukoulis, "Ministop bands in single-defect photonic crystal waveguides," *Phys. Rev. E*, vol. 64, p. 055603, 2001.

[26] A. Y. Petrov and M. Eich, "Zero dispersion at small group velocities in photonic crystal waveguides," *Appl. Phys. Lett.*, vol. 85, pp. 4866–4868, 2004.

[27] E. Dulkeith, S. J. McNab, and Y. A. Vlasov, "Mapping the optical properties of slab-type two-dimensional photonic crystal waveguides," *Phys. Rev. B*, vol. 72, p. 115102, 2005.

[28] Y. Tanaka, T. Asano, Y. Akahane, B. S. Song, and S. Noda, "Theoretical investigation of a two-dimensional photonic crystal slab with truncated cone air holes," *Appl. Phys. Lett.*, vol. 82, pp. 1661–1663, 2003.

[29] D. Mori and T. Baba, "Dispersion-controlled optical group delay device by chirped photonic crystal waveguides," *Appl. Phys. Lett.*, vol. 85, pp. 1101–1103, 2004.

[30] W. Kuang, C. Kim, A. Stapleton, and J. D. O'Brien, "Grating-assisted coupling of optical fibers and photonic crystal waveguides," *Opt. Lett.*, vol. 27, pp. 1604–1606, 2004.

[31] P. E. Barclay, K. Srinivasan, M. Borselli, and O. Painter, "Efficient input and output fiber coupling to a photonic crystal waveguide," *Opt. Lett.*, vol. 29, pp. 697–699, 2004.

[32] F. Van Laere, G. Roelkens, M. Ayre, J. Schrauwen, D. Taillaert, D. Van Thourhout, T. F. Krauss, and R. Baets, "Compact and highly efficient grating couplers between optical fiber and nanophotonic waveguides," *J. Lightw. Technol.*, vol. 25, pp. 151–156, 2007.

[33] V. R. Almeida, R. R. Panepucci, and M. Lipson, "Nanotaper for compact mode conversion," *Opt. Lett.*, vol. 28, pp. 1302–1304, 2003.

[34] S. J. McNab, N. Moll, and Y. A. Vlasov, "Ulatra-low loss photonic integrated circuit with membrane-type photonic crystal waveguides," *Opt. Express*, vol. 11, no. 22, pp. 2927–2939, 2003.

[35] A. Sure, T. Dillon, J. Murakowski, C. Lin, D. Pustai, and D. W. Prather, "Fabrication and characterization of three-dimensional silicon tapers," *Opt. Express*, vol. 11, no. 26, pp. 3555–3561, 2003.

[36] P. D. W., S. Shi, J. Murakowski, G. J. Schneider, A. Sharkawy, C. Chen, and B. Miao, "Photonic crystal structures and applications: Perspective, overview, and development," *IEEE J. Sel. Top. Quantum Electron.*, vol. 12, no. 6, pp. 1416–1436, 2006.

[37] S. G. Johnson, P. Bienstman, M. A. Skorobogatiy, M. Ibanescu, E. Lidorikis, and J. D. Joannopoulos, "Adiabatic theorem and continuous coupled-mode theory for efficient taper transitions in photonic crystals," *Phys. Rev. E*, vol. 66, p. 066608, 2002.

[38] A. Adibi, Y. Xu, R. K. Lee, A. Yariv, and A. Scherer, "Guiding mechanisms in dielectric-core photonic-crystal optical waveguides," *Phys. Rev. B*, vol. 64, p. 033308, 2001.

[39] A. Y. Petrov and M. Eich, "Dispersion compensation with photonic crystal line-defect waveguides," *IEEE J. Sel. Areas Comm.*, vol. 23, pp. 1396–1401, 2005.

[40] P. Pottier, M. Gnan, and R. M. De La Rue, "Efficient coupling into slow-light photonic crystal channel guides using photonic crystal tapers," *Opt. Express*, vol. 15, pp. 6569–6575, 2007.

[41] P. Dumon, W. Bogaerts, V. Wiaux, J. Wouters, S. Beckx, J. Van Campenhout, D. Taillaert, B. Luyssaert, P. Bienstmann, D. Van Thourhout, and R. Baets, "Low-loss SOI photonic wires and ring resonators fabricated with deep UV lithography," *IEEE Photon. Technol. Lett.*, vol. 16, no. 5, p. 1328, 2004.

[42] L. O'Faolain, X. Yuan, D. McIntyre, S. Thoms, H. Chong, R. M. De La Rue, and T. F. Krauss, "Low-loss propagation in photonic crystal waveguides," *Electron. Lett.*, vol. 42, p. 25, 2006.

[43] E. D. Palik, *Handbook of Optical Constants of Solids*. San Diego, CA: Academic Press, 1998.

[44] H. K. Tsang, C. S. Wong, T. K. Liang, I. E. Day, S. W. Roberts, A. Harpin, J. Drake, and M. Asghari, "Optical dispersion, two-photon absorption and self-phase modulation in silicon waveguides at 1.5μm wavelength," *Appl. Phys. Lett.*, vol. 80, no. 3, pp. 416–418, 2002.

[45] R. A. Soref and B. R. Bennett, "Electrooptical effects in silicon," *IEEE J. Quantum Electron.*, vol. 23, pp. 123–129, 1987.

[46] T. K. Liang and H. K. Tsang, "Role of free carriers from two-photon absorption in Raman amplification in silicon-on-insulator waveguides," *Appl. Phys. Lett.*, vol. 84, pp. 2745–2747, 2004.

[47] M. Loncar, D. Nedeljkovic, T. P. Pearsall, J. Vuckovic, A. Scherer, S. Kuchinsky, and D. Allan, "Experimental and theoretical confirmation of bloch-mode light propagation in planar photonic crystal waveguides," *Appl. Phys. Lett.*, vol. 80, pp. 1689–1691, 2002.

[48] P. Lalanne, "Electromagnetic analysis of photonic crystal waveguides operating above the light cone," *IEEE J. Quantum Electron.*, vol. 38, pp. 800–804, 2002.

[49] L. C. Andreani and M. Agio, "Intrinsic diffraction losses in photonic crystal waveguides with line defects," *Appl. Phys. Lett.*, vol. 82, pp. 2011–2013, 2003.

[50] R. Ferrini, R. Hourd, H. Benisty, M. Qiu, and J. Moosburger, "Radiation losses in planar photonic crystals: two-dimensional representation of hole depth and shape by an imaginary dielectric constant," *J. Opt. Soc. Am. B*, vol. 20, pp. 469–478, 2003.

[51] W. Bogaerts, P. Bienstamn, and R. Baets, "Scattering at sidewall roughness in photonic crystal slabs," *Opt. Lett.*, vol. 28, pp. 689–691, 2003.

[52] M. Povinelli, S. G. Johnson, E. Lidorikis, J. D. Joannopoulos, and M. Soljacic, "Effect of a photonic band gap on scattering from waveguide disorder," *Appl. Phys. Lett.*, vol. 84, pp. 3639–3641, 2004.

[53] T. N. Langtry, A. A. Asatryan, L. C. Botten, C. M. de Sterke, R. C. McPhedran, and P. A. Robinson, "Effects of disorder in two-dimensional photonic crystal waveguides," *Phys. Rev. E*, vol. 68, p. 026611, 2003.

[54] C. Poulton, M. Mueller, and W. Freude, "Scattering from sidewall deformations in photonic crystals," *J. Opt. Soc. Am. B*, vol. 22, pp. 1211–1220, 2005.

[55] Z. Y. Li, X. Zhang, and Z. Q. Zhang, "Disordered photonic crystals understood by a perturbation formalism," *Phys. Rev. B*, vol. 61, pp. 15738–15748, 2000.

[56] E. Lidorikis, M. M. Sigalas, E. N. Economou, and C. M. Soukoulis, "Gap deformation and classical wave localization in disordered two-dimensional photonic-band-gap materials," *Phys. Rev. B*, vol. 61, pp. 13458–13464, 2000.

[57] D. Gerace and L. C. Andreani, "Disorder-induced losses in photonic crystal waveguides with line defects," *Opt. Lett.*, vol. 29, pp. 1897–1900, 2004.

[58] S. Hughes, L. Ramunno, J. F. Young, and J. E. Sipe, "Extrinsic optical scattering loss in photonic crystal waveguides: Role of fabrication disorder and photon group velocity," *Phys. Rev. Lett.*, vol. 94, p. 033903, 2005.

[59] E. Kuramochi, M. Notomi, S. Hughes, A. Shinya, T. Watanabe, and L. Ramunno, "Disorder-induced scattering loss of line-defect waveguides in photonic crystal slabs," *Phys. Rev. B*, vol. 72, p. 161318, 2005.

[60] Y. A. Vlasov, M. O'Bolye, H. F. Hamann, and S. J. McNab, "Active control of slow light on a chip with photonic crystal waveguides," *Nature*, vol. 438, pp. 65–69, November 2005.

[61] L. Gu, W. Jiang, X. Chen, L. Wang, and R. T. Chen, "High speed silicon photonic crystal waveguide modulator for low voltage operation," *Appl. Phys. Lett.*, vol. 90, pp. 0711105–1–071105–3, February 2007.

[62] S. M. Weiss, H. Ouyang, J. Zhang, and P. M. Fauchet, "Electrical and thermal modulation of silicon photonic bandgap microcavities containing liquid crystals," *Opt. Express*, vol. 13, pp. 1090–1097, 2005.

[63] J. A. McCaulley, V. M. Donnelly, M. Vernon, and I. Taha, "Temperature dependence of the near-infrared refractive index of silicon, gallium arsenide, and indium phosphide," *Phys. Rev. B*, vol. 49, pp. 7408–7417, 1994.

[64] R. W. Boyd, *Nonlinear Optics*. San Diego: Academic Press, 2003.

[65] Y. Enami, C. T. Derose, D. Mathine, C. Loychik, C. Greenlee, R. A. Norwook, T. D. Kim, J. Luo, Y. Tian, A. K. Y. Jen, and N. Peyghambarian, "Hybrid polymer/so-gel waveguide modulators with exceptionally large electro-optic coefficients," *Nature Photonics*, vol. 1, pp. 180–185, 2007.

[66] R. W. Boyd, D. J. Gauthier, and A. L. Gaeta, "Applications of slow light in telecommunications," *Optics and Photonics News*, vol. 17, pp. 18–23, 2006.

[67] R. S. Tucker, P. C. Ku, and C. J. Chang-Hasnain, "Slow-light optical buffers: Capabilities and fundamental limitations," *J. Lightw. Technol.*, vol. 23, pp. 4046–4066, 2005.

[68] C. Liu, Z. Dutton, C. H. Behroozi, and L. V. Hau, "Observation of coherent optical information storage in an atomic medium using halted light pulses," *Nature*, vol. 409, pp. 490–493, 2001.

[69] D. Budker, D. F. Kimball, and V. V. Rochester, S. M. Yashchuk, "Nonlinear magneto-optics and reduced group velocity of light in atomic vapor with slow ground state relaxation," *Phys. Rev. Lett.*, vol. 83, pp. 1767–1770, 1999.

[70] A. V. Turukhin, V. S. Sudarshanam, M. S. Shahriar, J. A. Musser, B. S. Ham, and P. R. Hemmer, "Observation of ultraslow and stored light pulses in a solid," *Phys. Rev. Lett.*, vol. 88, p. 023602, 2002.

[71] M. S. Bigelow, N. N. Lepeshkin, and R. W. Boyd, "Observation of ultraslow light propagation in a ruby crystal at room temperature," *Phys. Rev. Lett.*, vol. 90, p. 113903, 2003.

[72] E. Baldit, E. Bencheikh, P. Monnier, J. A. Levenson, and V. Rouget, "Ultraslow light propagation in an inhomogeneously broadened rare-earth ion-doped crystal," *Phys. Rev. Lett.*, vol. 95, p. 143601, 2005.

[73] G. M. Gehring, A. Schweinsberg, C. Barsi, N. Kostinski, and R. W. Boyd, "Observation of backward pulse propagation through a medium with a negative group velocity," *Science*, vol. 312, pp. 895–897, 2005.

[74] X. Zhao, P. Palinginis, B. Pesala, C. J. Chang-Hasnain, and P. Hemmer, "Tunable ultraslow light in vertical-cavity surface-emitting laser amplifier," *Opt. Express*, vol. 13, pp. 7899–7904, 2005.

[75] P. C. Ku, F. Sedgwick, C. J. Chang-Hasnain, P. Palinginis, T. Li, H. Wang, S. W. Chang, and S. L. Chuang, "Slow light in semiconductor quantum wells," *Opt. Lett.*, vol. 29, pp. 2291–2293, 2004.

[76] J. Mork, R. Kjaer, M. van der Poel, and K. Yvind, "Slow light in a semiconductor waveguide at gigahertz frequencies," *Opt. Express*, vol. 13, pp. 8136–8145, 2005.

[77] Y. Okawachi, M. S. Bigelow, J. E. Sharping, Z. Zhu, A. Schweinsberg, D. J. Gauthier, R. W. Boyd, and A. L. Gaeta, "Tunable all-optical delays via Brillouin slow light in an optical fiber," *Phys. Rev. Lett.*, vol. 94, p. 153902, 2005.

[78] K. Y. Song, M. G. Herraez, and L. Thevenaz, "Observation of pulse delaying and advancement in optical fibers using stimulated Brillouin scattering," *Opt. Express*, vol. 13, pp. 82–88, 2005.

[79] J. Sharping, Y. Okawachi, and A. Gaeta, "Wide bandwidth slow light using a Raman fiber amplifier," *Opt. Express*, vol. 13, pp. 6092–6098, 2005.

[80] D. Dahan and G. Eisenstein, "Tunable all optical delay via slow and fast light propagation in a Raman assisted fiber optical parametric amplifier: a route to all optical buffering," *Opt. Express*, vol. 13, pp. 6234–6249, 2005.

[81] J. Sharping, Y. Okawachi, J. van Howe, C. Xu, Y. Wang, A. Willner, and A. Gaeta, "All-optical, wavelength and bandwidth preserving, pulse delay based on parametric wavelength conversion and dispersion," *Opt. Express* vol. 13, pp. 7872–7877, 2005.

[82] G. Lenz, B. J. Eggleton, C. K. Madsen, and R. E. Slusher, "Optical delay lines based on optical filters," *IEEE J. Quantum Electron.*, vol. 37, pp. 525–532, 2001.

[83] M. L. Povinelli, S. G. Johnson, and J. D. Joannopoulos, "Slow-light, band-edge waveguides for tunable time delays," *Opt. Express*, vol. 13, pp. 7145–7159, 2005.

[84] Q. Xu, P. Dong, and M. Lipson, "Breaking the delay-bandwidth limit in a photonic structure," *Nature Physics*, vol. 3, pp. 406–410, 2005.

[85] F. Xia, L. Sekaric, and Y. Vlasov, "Ultracompact optical buffers on a silicon chip," *Nature Photonics*, vol. 1, pp. 65–71, 2007.

[86] A. Yariv, Y. Xu, R. K. Lee, and A. Scherer, "Coupled-resonator optical waveguide: A proposal and analysis," *Opt. Lett.*, vol. 24, pp. 711–713, 1999.

[87] M. F. Yanik, W. Suh, Z. Wang, and S. Fan, "Stopping light in a waveguide with an all-optical analog of electromagnetically induced transparency," *Phys. Rev. Lett.*, vol. 93, p. 233903, 2004.

[88] H. Gersen, T. J. Karle, R. J. P. Engelen, W. Bogaerts, J. P. Korterik, N. F. van Hulst, T. F. Krauss, and L. Kuipers, "Real-space observation of ulstraslow light in photonic crystal waveguides," *Phys. Rev. Lett.*, vol. 94, p. 073903, 2005.

[89] C. J. M. Smith, H. Benisty, S. Olivier, M. Rattier, C. Weisbuch, T. F. Krauss, R. M. De La Rue, R. Houdre, and U. Oseterle, "Low-loss channel waveguides with two-dimensional photonic crystal boundaries," *Appl. Phys. Lett.*, vol. 77, pp. 2813–2815, 2000.

[90] H. Frandsen, L, A. Harpoth, P. I. Borel, M. Kristensen, S. S. Jensen, and O. Sigmund, "Broadband photonic crystal waveguide 60 bend obtained utilizing topology optimization," *Opt. Express*, vol. 12, pp. 5916–5921, 2004.

[91] Y. Watanabe, N. Ikeda, Y. Sugimoto, Y. Takata, Y. Kitagawa, A. Mizutani, N. Ozaki, and K. Asakawa, "Topology optimization of waveguide bends with wide, flat bandwidth in air-bridge-type photonic crystal slabs," *Journ. of Appl. Phys.*, vol. 10, p. 113108, 2007.

[92] R. S. Jacobsen, A. V. Lavrinenko, L. H. Frandsen, C. Peucheret, B. Zsigri, G. Moulin, J. Fage-Pedersen, and P. I. Borel, "Direct experimental and numerical determination of extremely high group indices in photonic crystal waveguides," *Opt. Express*, vol. 13, pp. 7861–7871, 2005.

[93] L. H. Frandsen, A. V. Lavrinenko, J. Fage-Pedersen, and P. I. Borel, "Photonic crystal waveguides with semi-slow light and tailored dispersion properties," *Opt. Express*, vol. 14, pp. 9444–9450, 2006.

[94] S. Kubo, D. Mori, and T. Baba, "Low-group-velocity and low-dispersion slow light in photonic crystal waveguides," *Opt. Lett.*, vol. 32, pp. 2981–2983, 2007.

[95] M. D. Settle, R. J. P. Engelen, M. Salib, A. Michaeli, L. Kuipers, and T. F. Krauss, "Flat-band slow light in photonic crystals featuring spatial pulse compression and terahertz bandwidth," *Opt. Express*, vol. 15, pp. 219–226, 2007.

[96] D. Mori, S. Kubo, H. Sasaki, and T. Baba, "Experimantal demonstration of wideband dispersion-compensated slow light by a chirped photonic crystal directional coupler," *Opt. Express*, vol. 15, pp. 5264–5270, 2007.

[97] M. Notomi, A. Shinya, K. Yamada, J. Takahashi, C. Takahashi, and I. Yokohama, "Structural tuning of guiding modes of line-defect waveguides of silicon-on-insulator photonic crystal slabs," *IEEE J. Quantum Electron.*, vol. 38, pp. 736–742, 2002.

[98] V. R. Almeida, Q. Xu, C. A. Barrios, and M. Lipson, "Guiding and confining light in void nanostructure," *Opt. Lett.*, vol. 29, no. 11, pp. 1209–1211, 2004.

[99] C. Koos, L. Jacome, C. Poulton, J. Leuthold, and W. Freude, "Nonlinear silicon-on-insulator waveguides for all-optical signal processing," *Opt. Express*, vol. 15, pp. 5976–1590, 2007.

[100] Y. A. Vlasov and S. J. McNab, "Coupling into the slow light mode in slab-type photonic crystal waveguides," *Opt. Lett.*, vol. 31, pp. 50–52, 2006.

[101] W. P. Huang, "Coupled-mode theory for optical waveguides: an overview," *J. Opt. Soc. Am.*, vol. 11, pp. 963–983, 1994.

[102] H. A. Haus, W. P. Huang, S. Kawakami, and N. A. Whitaker, "Coupled-mode theory of optical waveguides," *J. Lightw. Technol.*, vol. LT-5, pp. 16–23, 1987.

[103] L. O'Faolain, T. P. White, D. O'Brien, X. Yuan, M. D. Settle, and T. F. Krauss, "Dependence of extrinsic loss on group velocity in photonic crystal waveguides," *Opt. Express*, vol. 15, pp. 13129–13138, 2007.

[104] D. Marcuse, *Theory of Dielectric Optical Waveguides*. London: Academic Press, 1974.

[105] D. Marcuse, "Coupled power equations for backward waves," *IEEE Trans. Microwave Theory Tech.*, vol. MTT-20, pp. 541–546, 1972.

[106] B. S. Song, S. Noda, T. Asano, and Y. Akahane, "Ultra-high-Q photonic double-heterostructure nanocavity," *Nature Materials*, vol. 4, pp. 207–210, 2005.

[107] Y. Akahane, T. Asano, B. S. Song, and S. Noda, "High-Q photonic nanocavity in a two-dimensional photonic crystal," *Nature*, vol. 425, pp. 944–947, 2003.

[108] L. H. Frandsen, P. I. Borel, Y. X. Zhuang, A. Harpoth, M. Thorhauge, M. Kristensen, W. Bogaerts, P. Dumon, R. Baets, V. Wiaux, J. Wouters, and S. Beckx, "Ultralow-loss 3-dB photonic crystal waveguide splitter," *Opt. Lett.*, vol. 29, pp. 1623–1625, 2004.

[109] Y. Akahane, T. Asano, H. Takano, B. S. Song, Y. Takana, and S. Noda, "Two-dimensional photonic-crystal-slab channel-drop filter with flat-top response," *Opt. Express*, vol. 13, pp. 2512–2530, 2005.

[110] F. S. S. Chien, Y. J. Hsu, W. F. Hsieh, and S. C. Cheng, "Dual wavelength demultiplexing by coupling and decoupling of photonic crystal waveguides," *Opt. Express*, vol. 12, pp. 1119–1125, 2004.

[111] O. Painter, R. K. Lee, A. Scherer, A. Yariv, J. D. O'Brien, P. D. Dapkus, and I. Kim, "Two-dimensional photonic band-gap defect mode laser," *Science*, vol. 284, pp. 1819–1821, 1999.

[112] W. d. Zhou, J. Sabarinathan, B. Kochman, E. Berg, O. Qasaimeh, S. Pang, and P. Bhattacharya, "Electrically injected single-defect photonic bandgap surface-emitting laser at room temperature," *Electron. Lett.*, vol. 36, pp. 1541–1542, 2000.

[113] T. Tanabe, M. Notomi, S. Mitsugi, A. Shinya, and E. Kuramochi, "Fast bistable all-optical switch and memory on a silicon photonic crystal on-chip," *Opt. Lett.*, vol. 30, pp. 2575–2577, 2005.

[114] A. J. Lowery and J. Armstrong, "Orthogonal-frequency-division multiplexing for dispersion compensation of long-haul optical systems," *Opt. Express*, vol. 14, pp. 2079–2084, 2006.

[115] T. L. Koch and R. C. Alferness, "Dispersion compensation by active predistorted signal synthesis," *J. Lightw. Technol.*, vol. LT-3, pp. 800–805, 1985.

[116] J. H. Winters and R. D. Gitlin, "Electrical signal processing techniques in long-haul fiber-optic systems," *IEEE Trans. Commun.*, vol. 38, pp. 1439–1453, 1990.

[117] R. I. Killey, P. M. Watts, V. Mikhailov, M. Glick, and P. Bayvel, "Electronic dispersion compensation by signal predistortion using digital processing and a dual-drive MachZehnder modulator," *IEEE Photon. Technol. Lett.*, vol. 17, pp. 714–716, 2005.

[118] J. X. Cai, K. M. Feng, A. E. Willner, V. Grubsky, D. S. Starodubov, and J. Feinberg, "Simultaneous tunable dispersion compensation of many WDM channels using a sampled nonlinearly chirped fiber Bragg grating," *IEEE Photon. Technol. Lett.*, vol. 11, pp. 1455–1457, 1999.

[119] M. Shirasaki, "Chromatic-dispersion compensator using virtually imaged phased array," *IEEE Photon. Technol. Lett.*, vol. 9, pp. 1598–1600, 1997.

[120] D. J. Moss, M. Lamont, S. McLaughlin, G. Randall, P. Colbourne, S. Kiran, and C. A. Hulse, "Tunable dispersion and dispersion slope compensators for 10 Gb/s using all-pass multicavity etalons," *IEEE Photon. Technol. Lett.*, vol. 15, pp. 730–732, 2003.

[121] C. K. Madsen, J. A. Walker, K. W. Goossen, T. N. Nielsen, and G. Lenz, "A tunable dispersion compensating MEMS all-pass filter," *IEEE Photon. Technol. Lett.*, vol. 12, pp. 651–653, 2000.

[122] U. Peschel, T. Peschel, and F. Lederer, "A compact device for highly efficient dispersion compensation in fiber transmission," *Appl. Phys. Lett.*, vol. 67, pp. 2111–2113, 1995.

[123] C. K. Madsen, G. Lenz, A. J. Bruce, M. A. Cappuzzo, L. T. Gomez, and R. E. R. E. Scotti, "Integrated all-pass filters for tunable dispersion and dispersion slope compensation," *IEEE Photon. Technol. Lett.*, vol. 11, pp. 1623–1625, 1999.

[124] K. Takiguchi, K. Jinguji, K. Okamoto, and Y. Ohmori, "Variable group-delay dispersion equalizer using lattice-form programmable optical filter on planar lightwave circuit," *IEEE J. Sel. Top. Quantum Electron.*, vol. 2, pp. 270–276, 1996.

[125] C. R. Doerr, L. W. Stulz, S. Chandrasekhar, and R. Pafchek, "Colorless tunable dispersion compensator with 400-ps/nm range integrated with a tunable noise filter," *IEEE Photon. Technol. Lett.*, vol. 15, pp. 1258–1260, 2003.

[126] A. Yariv, D. Fekete, and D. M. Pepper, "Compensation for channel dispersion by nonlinear optical phase conjugation," *Opt. Lett.*, vol. 4, pp. 52–54, 1979.

[127] D. Mogilevtsev, T. A. Birks, and P. S. J. Russell, "Group-velocity dispersion in photonic crystal fibers," *Opt. Lett.*, vol. 23, pp. 1662–1664, 1998.

[128] A. Y. Petrov and M. Eich, "Efficient approximation to calculate time delay and dispersion in linearly chirped periodical microphotonic structures," *IEEE J. Quantum Electron.*, vol. 41, pp. 1502–1509, 2005.

[129] W. J. Kim, W. Kuang, and J. D. OBrien, "Dispersion characteristics of photonic crystal coupled resonator optical waveguides," *Opt. Express*, vol. 11, pp. 3431–3437, 2003.

[130] Z. Zhu, A. M. C. Dawes, D. J. Gauthier, L. Zhang, and A. E. Willner, "Broadband SBS slow light in an optical fiber," *J. Lightw. Technol.*, vol. 25, pp. 201–206, 2007.

[131] Y. Okawachi, R. Salem, and A. L. Gaeta, "Continuous tunable delays at 10-Gb/s data rates using self-phase modulation and dispersion," *J. Lightw. Technol.*, vol. 25, pp. 3710–3715, 2007.

[132] F. Morichetti, A. Melloni, A. Breda, A. Canciamilla, C. Ferrari, and M. Martinelli, "A reconfigurable architecture for continuously variable optical slow-wave delay lines," *Opt. Express*, vol. 15, pp. 17273–17282, 2007.

[133] L. Liao, A. Liu, D. Rubin, J. Basak, Y. Chetrit, H. Nguyen, R. Cohen, N. Izhaky, and M. Paniccia, "40Gbit/s silicon optical modulator for high-speed applications," *Electron. Lett.*, vol. 43, p. 20072253, 2007.

[134] B. Bortnik, Y. C. Hung, H. Tazawa, B. J. Seo, J. Luo, A. K. Y. Jen, W. H. Steier, and H. R. Fetterman, "Electrooptic polymer ring resonator modulation up to 165GHz," *IEEE J. Sel. Top. Quantum Electron.*, vol. 13, pp. 104–110, 2007.

[135] D. Rezzonico, M. Jazbinsek, A. Guarino, O. P. Kwon, and P. Günter, "Electro-optic charon polymeric microring modulators," *Opt. Express*, vol. 16, pp. 613–627, 2008.

[136] E. M. McKenna, A. S. Lin, A. R. Mickelson, R. Dinu, and D. Jin, "Comparison of r_{33} values for AJ404 films prepared with parallel plate and corona poling," *J. Opt. Soc. Am. B*, vol. 24, pp. 2888–2892, 2007.

[137] B. Wang, T. Baehr-Jones, M. Hochberg, and A. Scherer, "Design and fabrication of segmented, slotted waveguides for electro-optic modulation," *Appl. Phys. Lett.*, vol. 91, p. 143109, 2007.

[138] Q. Xu, B. Schmidt, S. Pradhan, and M. Lipson, "Micrometre-scale silicon electro-optic modulator," *Nature*, vol. 435, pp. 325–327, 2005.

[139] B. Schmidt, Q. Xu, J. Shakya, S. Manipatruni, and M. Lipson, "Compact electro-optic modulator on silicon-on-insulator substrates using cavities with ultra-small modal volumes," *Opt. Express*, vol. 15, pp. 3140–3148, 2007.

[140] K. K. McLauchlan and S. T. Dunham, "Analysis of a compact modulator incorporating a hybrid silicon/electro-optic polymer waveguide," *IEEE J. Sel. Top. Quantum Electron.*, vol. 12, pp. 1455–1460, 2006.

[141] L. C. Andreani and D. Gerace, "Photonic-crystal slabs with a triangular lattice of triangular holes investigated using a guided-mode expansion method," *Phys. Rev. B*, vol. 73, p. 235114, 2006.

[142] C. Koos, *Nanophotonic Devices for Linear and Nonlinear Optical Signal Processing*. PhD thesis, Universität Karlsruhe (TH), 2008.

[143] M. Notomi, A. Shinya, S. Mitsugi, E. Kuramochi, and H. Y. Ryu, "Waveguides, resonators and their coupled elements in photonic crystal slabs," *Opt. Express*, vol. 12, pp. 1551–1561, 2004.

[144] W. Bogaerts, R. Baets, P. Dumon, V. Wiaux, S. Beckx, D. Taillaert, B. Luyssaert, J. van Campenhout, P. Bienstman, and D. van Thourhout, "Nanophotonic waveguides in silicon-on-insulator fabricated with CMOS technology," *J. Lightw. Technol.*, vol. 23, pp. 401–412, 2005.

[145] A. D. Falco, L. OFaolain, and T. F. Krauss, "Dispersion control and slow light in slotted photonic crystal waveguides," *Appl. Phys. Lett.*, vol. 92, p. 083501, 2008.

[146] W. Bogaerts, R. Baets, P. Dumon, V. Wiaux, S. Beckx, D. Taillaert, B. Luyssaert, J. Van Campenhout, P. Bienstmann, and D. Van Thourhout, "Nanophotonic waveguides in silicon-on-insulator fabricated with CMOS technology," *J. Lightwave Technol.*, vol. 23, no. 1, pp. 401–412, 2005.

[147] J. Canning, N. Skivesen, M. Kristensen, L. H. Frandsen, A. Lavrinenko, C. Martelli, and A. Tetu, "Mapping the broadband polarization properties of linear 2D SOI photonic crystal waveguides," *Opt. Express*, vol. 15, pp. 15603–15614, 2007.

[148] S. L. McCall, P. M. Platzmann, R. Dalichaouch, D. Smith, and S. Schultz, "Microwave propagation in two-dimensional dielectric lattices," *Phys. Rev. Lett.*, vol. 67, pp. 2017–2020, 1991.

[149] C. Jin, b. Cheng, B. Man, Z. Li, and D. Zhang, "Two-dimensional dodecagonal and decagonal quasiperiodic photonic crystals in the microwave region," *Phys. Rev. B*, vol. 61, p. 10762, 2000.

[150] D. R. Solli, C. F. McCormick, R. Y. Chiao, and J. M. Hickmann, "Experimental demonstration of photonic crystal waveplates," *Appl. Phys. Lett.*, vol. 82, pp. 1036–1038, 2003.

[151] E. Ozbay, M. Bayindir, I. Bulu, and E. Cubukcu, "Investigation of localized coupled-cavity modes in two-dimensional photonic bandgap structures," *IEEE J. Quantum Electron.*, vol. 38, pp. 837–843, 2002.

[152] F. Cuesta, A. Griol, A. Martinez, and J. Marti, "Experimental demonstration of photonic crystal directional coupler at microwave frequencies," *Electron. Lett.*, vol. 39, pp. 455–456, 2003.

[153] A. A. Chabanov, M. Stoytchev, and A. Z. Genack, "Statistical signatures of photon localization," *Nature*, vol. 404, pp. 850–853, 2000.

[154] R. A. Shelby, D. R. Smith, and S. Schultz, "Experimental verification of a negative index of refraction," *Science*, vol. 292, pp. 77–79, 2001.

[155] E. Cubukcu, K. Aydin, E. Ozbay, S. Foteinopoulou, and C. M. Soukoulis, "Negative refraction by photonic crystals," *Nature*, vol. 423, pp. 604–605, 2003.

[156] P. V. Parimi, W. T. Lu, P. Vodo, J. Sokoloff, J. S. Derov, and S. Sridhar, "Negative refraction and left-handed electromagnetism in microwave photonic crystals," *Phys. Rev. Lett.*, vol. 92, p. 127401, 2004.

[157] Z. Lu, J. A. Murakowski, C. A. Schuetz, S. Shi, G. J. Schneider, and D. W. Prather, "Three-dimensional subwavelength imaging by a photonic-crystal flat lens using negative refraction at microwave frequencies," *Phys. Rev. Lett.*, vol. 95, p. 153901, 2005.

[158] T. J. Karle, Y. J. Chai, C. N. Morgan, I. H. White, and T. F. Krauss, "Observation of pulse compression in photonic crystal coupled cavity waveguides," *J. Lightw. Technol.*, vol. 22, pp. 514–519, 2004.

[159] D. Mori and T. Baba, "Wideband and low dispersion slow light by chirped photonic crystal coupled waveguide," *Opt. Express*, vol. 13, pp. 9398–9408, 2005.

[160] CST Microwave Studio, http://www.cst.com/.

[161] RSOFT FullWAVE, http://www.rsoftdesign.com/.

[162] MIT photonic bands, http://ab-initio.mit.edu/wiki/index.php/MIT_Photonic_Bands/.

[163] Guided mode expansion, http://fisicavolta.unipv.it/dipartimento/ricerca/Fotonici/.

[164] G. Lecamp, J. P. Hugonin, and P. Lalanne, "Theoretical and computational concepts for periodic optical waveguides," *Opt. Express*, vol. 15, pp. 11052–11060, 2007.

[165] G. F. Engen and C. A. Hoer, "Thru-reflect-line: An improved technique for calibrating the dual six-port automatic network analyzer," *IEEE Trans. Microwave Theory Tech.*, vol. MTT-27, pp. 987–993, 1979.

Acknowledgments

This thesis is the result of my work at the Institute of High-Frequency and Quantum Electronics (IHQ) at the University of Karlsruhe (TH), Germany. We acknowledge the support by the Deutsche Forschungsgemeinschaft (DFG) in the framework of the Priority Program SP 1113 "Photonic Crystals", by the Deutsche Telekom Stiftung, and by the European FP6 Network of Excellence ePIXnet.

At this point I also want to thank all the people who have helped in making this work a success.

First of all, I want to thank my supervisors Prof. Wolfgang Freude and Prof. Jürg Leuthold. They have always supported me, brought forward inspiring ideas, and have given me plenty of room for independant work. My first supervisor, Prof. Wolfgang Freude, I want to thank for his great analytic expertise that always helped to get technical issues to the point and find solutions, his thoroughness that prevented mistakes to creep in, and for his great helpfulness and his always open door where I could step in at any time. Our institute head, Prof. Jürg Leuthold, I especially want to thank for his vision that helped to see the big picture and his steadfast optimism to bring small commencements finally to a success.

Prof. Lucio Claudio Andreani of the University of Pavia, Italy, I want to thank for the collaboration in the simulation of band diagrams and of disorder-induced losses with the guided mode expansion (GME) method. He provided and advanced his GME code for our applications, and many helpful discussions took place.

Dr. Alexander Petrov, University of Hamburg-Harburg, I thank for the collaboration and discussion on broadband slow-light photonic crystal waveguides and for his help in the first design that we realized in the microwave model.

Dr. Pieter Dumon of the Interuniversity Microelectronics Center (IMEC), Leuven, Belgium, I thank for the help in the mask design and for his support in the successful fabrication of our silicon-on-insulator (SOI) devices in the ePIXnet framework.

Peter Pfundstein of the Laboratory for Microscopy (LEM), University of Karlsruhe, I want to thank for conducting the scanning electron microscope pictures of our SOI samples.

Prof. Uli Lemmer and Christian Kayser of the Light Technology Institute (LTI), University of Karlsruhe, I thank for giving access to their clean room and for the instructions on spin-coating.

My former colleague Dr. Christian Koos I especially thank for the countless helpful discussions in numerous fields, for the collaboration in the SOI design and the provision of his characterization setup, and for his deep knowledge he was always willing to share. My former colleague Dr. Christopher Poulton I thank for the discussions on the fundamentals of photonic crystals and of disorder effects, and for the collaboration on numerical modeling.

The mechanical workshop, Hans Bürger, Manfred Hirsch, Werner Höhne, Oswald Speck and the apprentices I thank for the fabrication of the microwave model structures and for the execution of all kinds of other technical tasks. Oswald Speck of the packaging lab I thank for the great support in wafer cleaving, lapping and handling.

Andrej Marculescu I thank for his extensive help in all kinds of computer issues, and for the provision of optical measurement equipment. Philipp Vorreau I thank for the discussions about and the provision of microwave and optical measurement equipment. Thomas Vallaitis I want to thank for the discussions concerning silicon photonics and semiconductors.

Many thanks to all the students who have contributed to our work: Michael Waldow, Mireia Monllor Tormos, Irene Netsch, Simon Schneider and Christoph Heine.

For the great collaboration I want to thank all my colleagues at IHQ, René Bonk, Dr. Gunnar Böttger, Dr. Shunfeng Li, Jingshi Li, Andrej Marculescu, Moritz Röger, Dr. Stelios Sygletos, Thomas Vallaitis, Philipp Vorreau, and Jin Wang and my former colleagues Dr. Guy-Aimar Chakam, Dr. Richard Ell, Anastasiya Golovina, Dr. Martin Hugelmann, Dr. Philipp Hugelmann, Lenin Jacome, Dr. Christian Koos, Dr. Ayan Maitra, Dr. Christopher Poulton, Dr. Wolfgang Seitz, Oleksiy Shulika and Dr. Philipp Wagenblast.

Many thanks to our secretaries, Bernadette Lehmann and Dagmar Goldmann, for the help in administrative matters, and to Ilse Kober for the technical artwork. Thanks to Peter Frey and Werner Podszus for the support in electrical and technical matters.

Besides the work-related concerns, I also want to thank all my colleagues for the personal friendships, for the good working climate, for the encouragement, and for the times of relaxation and fun. Moritz and Gunnar, it is a pleasure to share the office with you.

Many thanks to my parents Willi and Heidemarie Brosi and to my family for their support. I also want to thank my girlfriend Christiane Sutter for all her help, her care and her affection.

Finally, I thank my Saviour Jesus, who is the reason of my life. "For from him and through him and to him are all things. To him be the glory for ever." (Romans 11:36, the Bible)

List of Own Publications

Patents

P1 Koos, C.; Leuthold, J.; Freude, W., **Brosi, J.-M.**: *Elektro-optisches Bauteil mit transparenten Elektroden, kleiner Betriebsspannung und hoher Grenzfrequenz*, file numbers DE 10 2006 045 102 A1, WO 2008/034559 A1

Journal Papers

J3 **Brosi J.-M.**; Koos C.; Andreani L. C.; Waldow M.; Leuthold J.; Freude W.: *High-speed low-voltage electro-optic modulator with a polymer-infiltrated silicon photonic crystal waveguide*, Opt. Expr. 16 (2008) 4177–4191

J2 **Brosi J.-M.**; Leuthold J.; Freude W.: *Microwave-frequency experiments validate optical simulation tools and demonstrate novel dispersion-tailored photonic crystal waveguides*, J. Lightw. Technol. 25 (2007) 2502–2510

J1 Stalzer H.; Cosceev A.; Sürgers C.; v. Löhneysen H.; **Brosi J.-M.**; Chakam G.-A.; Freude W.: *Inhomogeneous magnetization of a superconducting film measured with a gradiometer*, Appl. Phys. Lett. 84 (2004) 1522–1524

Book Chapter

B1 Freude, W.; Chakam, G.-A.; **Brosi, J.-M.**; Koos, C.: *Microwave Modelling of Photonic Crystals*. In: Busch, K.; Lölkes, S.; Wehrspohn, R.; Föll, H. (Eds.): *Photonic Crystals – Advances in Design, Fabrication, and Characterization*. Wiley VCH, Berlin 2004, pp.198–214

Conference Contributions

C20 **Brosi J.-M.**; Koos C.; Andreani L. C.; Dumon P.; Baets R.; Leuthold J.; Freude W.: *100 Gbit/s / 1 V Optical Modulator With Slotted Slow-Light Polymer-Infiltrated Silicon Photonic Crystal*. Slow and Fast Light, OSA Topical Meeting, Boston MA, USA; July 13-16, 2008

C19 Freude W.; **Brosi J.-M.**; Koos C.; Vorreau P.; Andreani L. C.; Dumon P.; Baets R.; Esembeson B.; Biaggio I.; Michinobu T.; Diederich F.; Leuthold J.: *Silicon-organic hybrid (SOH) devices for nonlinear optical signal processing.* Proc. 10th Intern. Conf. on Transparent Optical Networks (ICTON'08), June 22–26, 2008, Athens, Greece **(invited)**

C18 **Brosi J.-M.**; Freude W.; Leuthold J.: *Schneller elektro-optischer Modulator als silizium-organische Hybridstruktur auf der Basis von photonischen Kristallen.* VDE-ITG Workshop der Fachgruppe 5.3.1, Kiel, Germany, June 12–13, 2008.

C17 Freude W.; **Brosi J.-M.**; Koos C.; Vorreau P.; Andreani L. C.; Dumon P.; Baets R.; Esembeson B.; Biaggio I.; Michinobu T.; Diederich F.; Leuthold J.: *Silicon photonics meets nonlinear organic materials.* Proc. COST-MP0702 Workshop and WG Meeting, Warsaw, Poland, April 28–29, 2008. Paper P2.3.1, pp. 60–62

C16 Koos C.; **Brosi J.**; Waldow M.; Freude W.; Leuthold J.: *Silicon-on-insulator modulators for next-generation 100 Gbit/s-Ethernet.* Proc. 33th European Conf. Opt. Commun. (ECOC'07), Berlin, Germany; September 16–20, 2007, Paper P056

C15 Böttger G.; **Brosi J.-M.**; Maitra A.; Wang J.; Petrov A.Y.; Eich M.; Leuthold J.; Freude, W.: *Broadband slow light and nonlinear switching devices.* Proc. Progress in Electromagnetics Research Symposium 2006 (PIERS'07), Beijing, China; March 26–30, 2007, pp.1338-1342 **(invited)**

C14 Freude, W.; Fujii, M.; Koos, C.; **Brosi, J.**; Poulton, C. G.; Wang, J.; Leuthold, J.: *Wavelet FDTD methods and applications in nano-photonics*, 386th WE Heraeus Seminar 'Computational Nano-Photonics', Bad Honnef, February 25–28, 2007 **(invited)**

C13 **Brosi J.-M.**, Freude W., Leuthold J.: *Broadband Slow Light in a Photonic Crystal Line Defect Waveguide.* Proc. Slow and Fast Light, OSA Topical Meeting, OSA Headquarters, Washington DC, USA; July 23-26, 2006

C12 Freude W.; **Brosi J.**: *Finite-difference time-domain and beam propagation methods for Maxwell's equations.* 4-hours lecture with hands-on demonstrations given as part of the COST P11 Training School 'Modelling and Simulation Techniques for Linear, Nonlinear and Active Photonic Crystals'; University of Nottingham, UK; June, 2006

C11 Freude, W.; Poulton, C. G.; Koos, C.; **Brosi, J.**; Fujii, M.; Pfrang, A.; Schimmel, Th.; Müller, M.; Leuthold, J.: *Nanostrip and photonic crystal waveguides: The modelling of imperfections*, Proc. 15th International Workshop on Optical Waveguide Theory and Numerical Modelling (OWTNM'06), Varese, Italy, April 20–21, 2006, pp. 1–2 **(invited)**

C10 Freude W.; Poulton C.; Koos C.; **Brosi J.**; Fujii M.; Müller M.; Leuthold J.; Pfrang A.; Schimmel Th.: *Nano-optical waveguide devices with random deformations and nonlinearities.* Proc. 3rd Intern. Conf. on Materials for Advanced Technologies (ICMAT'05), Symposium F: Nano-Optics and Microsystems, Singapore; July 3–8, 2005, Paper F-2-N3 **(invited)**

C9 **Brosi J.**; Poulton C.; Freude W.: *Microwave modelling of disorder in photonic crystal slab waveguides.* Proc. Spring Meeting, Deutsche Physikalische Gesellschaft, DPG Quantum Optics Section, Berlin, Germany; March 2005

C8 Freude, W.; Poulton, C.; **Brosi, J.**; Koos, C.; Glöckler, F.; Wang, J.; Fujii, M.: *Integrated optical waveguide devices: Design, modeling and fabrication*, Proc. 16th Asia Pacific Microwave Conference (APMC'04), New Delhi, India, December 15–18, 2004 **(invited)**

C7 Freude, W.; **Brosi, J.**; Glöckler, F.; Koos, C.; Poulton, C.; Wang, J.; Fujii, M.: *High-index optical waveguiding structures*, Conf. Proc. Optical Society of America Annual Meeting (OSA 2004), Rochester (NY), USA, October 10–14, 2004, paper TuS1 **(invited)**

C6 Freude, W.; Poulton, C.; Koos, C.; **Brosi, J.**; Glöckler, F.; Wang, J.; Chakam, G.-A.; Fujii, M.: *Design and fabrication of nanophotonic devices*, Proc. 6th Intern. Conf. on Transparent Optical Networks (ICTON'04), July 4–8, 2004, Wroclaw, Poland. Mo.A.4 vol. 1 pp. 4–9 **(invited)**

C5 **Brosi, J.**; Chakam, G.-A.; Poulton, C.; Freude, W.: *Microwave Modeling of Integrated Optical Devices.* 323th WE Heraeus Seminar 'From Photonic Crystals to Metamaterials - Artificial Materials in Optics', Bad Honnef, April 26–30, 2004

C4 Poulton C.; **Brosi J.**; Koos C.; Glöckler F.; Wang J.; Fujii M.; Freude W.: *Simulation, Design and Fabrication of Integrated Optical Devices.* Proc. Third Joint Symposium on Opto- and Microelectronic Devices and Circuits (SODC'04), Wuhan, China; March 21–26, 2004, pp. 33–35

C3 Stalzer, H.; Cosceev, A.; Sürgers, C.; v. Löhneysen, H.; **Brosi, J.-M.**; Chakam, G.-A.; Freude, W.: *Proximity-Effekt in supraleitenden Nb/Ag- und Nb/Ag/Fe-Schichtpaketen.* Spring Meeting of the German Physical Society, Regensburg, March 8–12, 2004

C2 Freude, W.; **Brosi, J.-M.**; Chakam, G.-A.; Glöckler, F.; Koos, C.; Poulton, C.; Wang, J.: *Modelling and design of nanophotonic devices*, URSI Kleinheubacher Tagung, Miltenberg, Germany; September/October 2003 **(invited)**

C1 Poulton, C.; **Brosi, J.**; Glöckler, F.; Koos, C.; Wang, J.; Freude, W.; Ilin, K.; Siegel M.; Fujii, M.: *Modelling, design and realisation of SOI-based nanophotonic devices*; Conference on Functional Nanostructures 2003 (CFN03), Karlsruhe, Germany. September/October 2003

Curriculum Vitae

Jan-Michael Brosi

born March 21, 1978
in Filderstadt, Germany
Citizenship: German

Education

04/03–8/08 Dr.-Ing., Electrical Engineering, Institute of High-Frequency and Quantum Electronics (IHQ), University of Karlsruhe, Germany.

08/01–08/02 M.Sc., Electrical Engineering, Georgia Institute of Technology, Atlanta, USA. Specialization in Optics and Photonics, Microwave Engineering.

10/98–03/03 Studies of Electrical Engineering, University of Karlsruhe, Germany. Specialization in Optical Telecommunications.

Experience

08/02–10/02 Internship at Lucent Technologies, Nürnberg, Germany. Work on polarization mode dispersion in optical fibers.

01/02–8/02 Teaching assistant for course Electromagnetics, Georgia Institute of Technology, Atlanta, USA.

10/99–02/01 Teaching assistant for courses Electrophysics and Electronic Circuits, University of Karlsruhe, Germany.

07/98–09/98 Internship at Kolbenschmidt, Neckarsulm, Germany.
Training in the mechanical workshop and in various electronic labs.

07/97–06/98 Civillian Service, District hospital Bad Friedrichshall, Germany.

Scholarships and Awards

02/2005 Ph.D. scholarship of the Deutsche Telekom Stiftung (3 years).

05/2001 Study scholarship of Nortel Networks (3 years).

01/2001 Fulbright travel grant for the Georgia Institute of Technology, Atlanta.

10/2000 Scholarship for the exchange program with the Georgia Institute of Technology, Atlanta.

09/2000 Pre-degree award of the faculty of electrical engineering of the University of Karlsruhe